EXPLORATORY GALOIS THEORY

Combining a concrete perspective with an exploration-based approach, *Exploratory Galois Theory* develops Galois theory at an entirely undergraduate level. The text grounds the presentation in the concept of algebraic numbers with complex approximations and assumes of its readers only a first course in abstract algebra. The author organizes the theory around natural questions about algebraic numbers, and exercises with hints and proof sketches encourage students' participation in the development. For readers with *Maple* or *Mathematica*, the text introduces tools for hands-on experimentation with finite extensions of the rational numbers, enabling a familiarity never before available to students of the subject. *Exploratory Galois Theory* includes classical applications, from ruler-and-compass constructions to solvability by radicals, and also outlines the generalization from subfields of the complex numbers to arbitrary fields. The text is appropriate for traditional lecture courses, for seminars, or for self-paced independent study by undergraduates and graduate students.

John Swallow is J. T. Kimbrough Associate Professor of Mathematics at Davidson College. He holds a doctorate from Yale University for his work in Galois theory. He is the author or co-author of a dozen articles, including an essay in *The American Scholar*. His work has been supported by the National Science Foundation, the National Security Agency, and the Associated Colleges of the South.

Exploratory
Galois Theory

JOHN SWALLOW
Davidson College

CAMBRIDGE
UNIVERSITY PRESS

CAMBRIDGE
UNIVERSITY PRESS

32 Avenue of the Americas, New York NY 10013-2473, USA

Cambridge University Press is part of the University of Cambridge.

It furthers the University's mission by disseminating knowledge in the pursuit of education, learning and research at the highest international levels of excellence.

www.cambridge.org
Information on this title: www.cambridge.org/9780521544993

First published 2004

A catalogue record for this publication is available from the British Library

Mathematica® is a registered trademark of Wolfram Research. *Maple*™ is a registered trademark of Waterloo Maple, Inc.
Library of Congress Cataloguing in Publication data
Swallow, John, 1970–
        Exploratory Galois theory / John Swallow.
            p.   cm.
        Includes bibliographical references and index.
        ISBN 0-521-83650-6 (hardback) – ISBN 0-521-54499-8 (pbk.)
        1. Galois theory.   I. Title.
QA214.S93   2004
512′.32 – dc22   2004045196

ISBN 978-0-521-54499-3 Paperback

to Cameron

Contents

Preface

My goal in this text is to develop Galois theory in as accessible a manner as possible for an undergraduate audience.

Consequently, algebraic numbers and their minimal polynomials, objects as concrete as any in field theory, are the central concepts throughout most of the presentation. Moreover, the choices of theorems, their proofs, and (where possible) their order were determined by asking natural questions about algebraic numbers and the field extensions they generate, rather than by asking how Galois theory might be presented with utmost efficiency. Some results are deliberately proved in a less general context than is possible so that readers have ample opportunities to engage the material with exercises. In order that the development of the theory does not rely too much on the mathematical expertise of the reader, hints or proof sketches are provided for a variety of problems.

The text assumes that readers will have followed a first course in abstract algebra, having learned basic results about groups and rings from one of several standard undergraduate texts. Readers do not, however, need to know many results about fields. After some preliminaries in the first chapter, giving readers a common foundation for approaching the subject, the exposition moves slowly and directly toward the Galois theory of finite extensions of the rational numbers. The focus on the early chapters, in particular, is on building intuition about algebraic numbers and algebraic field extensions.

All of us build intuition by experimenting with concrete examples, and the text incorporates, in both examples and exercises, technological tools enabling a sustained exploration of algebraic numbers. These tools assist the exposition in proceeding with a concrete, constructive perspective, and, adopting this point of view, the text presents a Galois theory balanced between theory and computation. The exposition does not, however, fundamentally require or depend on these technological tools, and the text may usefully serve as a balanced introduction to Galois theory even for those who skip the computational sections and exercises.

The particular tools used in the text are contained in AlgFields, a package of functions designed for the symbolic computation systems *Maple* and *Mathematica*. This package is freely available for educational use at the website

http://www.davidson.edu/math/swallow/AlgFieldsWeb/index.htm.

The functions are introduced and explained in the occasional sections on computation. These sections treat both *Maple* and *Mathematica* at the same time, since, in general, only minor differences distinguish the syntax of the AlgFields package for the two symbolic computation systems. The text uses two-column displays to show input and output, *Maple* on the left and *Mathematica* on the right. (Line breaks are frequently inserted to facilitate the division of the page.) Now just as the text is not a comprehensive treatment of the Galois theory of arbitrary fields, sufficient for preparation for a qualifying exam in algebra in a doctoral program, the routines accompanying the text are not meant to display efficient algorithms for the determination of Galois groups and subfields of field extensions. Instead, the functions provide the ability to ask basic questions about algebraic numbers and to answer these questions using the very same methods and algorithms that appear in the theoretical exposition.

Despite the pedagogical use of computation in the early chapters, by the end of the text, students will be able to place what they have learned from a concrete study of algebraic numbers into a broader context of field theory in characteristic zero. In a pause before the Galois correspondence, the end of the fourth chapter introduces the general concepts of simple, algebraic, and finite extensions and explores the relations among these three properties. After a presentation of the Galois correspondence in the fifth chapter, the text also briefly treats various classical topics in Chapter 6, including cyclic extensions, binomial polynomials, ruler-and-compass constructions, and solvability by radicals.

For those readers for whom this text will be a jumping-off point for a deeper study of Galois theory, the necessary ideas and results for understanding the Galois theory of arbitrary fields are introduced in the penultimate section. That section contains problems leading readers to review previous results in light of a different perspective, one built not on concrete algebraic numbers with complex approximations but on isomorphism classes of arbitrary field extensions. Working through that section, these readers will gain the skills to approach later, with appreciation for the nuances, a more advanced and concise presentation of the subject. Furthermore, even without doing the exercises in that section, readers may profitably apply some of these ideas to finite fields, which are introduced in the final section.

This text may be effectively used in undergraduate major curricula as a second course in abstract algebra, and it would also serve well as a useful guide for a reading course or independent study. The text begins by presenting some standard results on fields in a basic fashion; depending on the content of the reader's first course in abstract algebra, the first chapter and the beginning of the second may be covered quickly. On the other hand, the text ends with a more challenging style, presenting in the last chapter some slightly abbreviated proofs with fewer references to prior theorems. These sections would be suitable for independent work by students in preparation for a class presentation. In fact, the entire text might productively be used for a seminar consisting of a group of students who learn to present this material to their peers; at Davidson College, I have used this material primarily in this fashion.

I would like to thank my many wonderful students at Davidson who have borne the burden of reading various drafts of this text and who have offered so many useful suggestions along the way. These students, Melanie Albert, Sandy Bishop, Frank Chemotti, Brent Dennis, Will Herring, Anders Kaseorg, Margaret Latterner, Chris Lee, Rebecca Montague, Dave Parker, Martha Peed, Joe Rusinko, Andy Schultz, and Ed Tanner, a group who at the time of writing this preface spend their time variously as doctors, graduate students, programmers, teachers, and ultimate players, helped to shape these materials while sharing with me their joy in learning a beautiful subject. I am moreover indebted to Nat Thiem, a coauthor and former summer research student, for his insights as a current graduate student in mathematics.

I also wish to express my great appreciation for friends and colleagues Jorge Aarão, Irl Bivens, Joe Gallian, David Leep, Ján Mináč, Pat Morandi, and Tara Smith, as well as for two reviewers whose names I do not know, for their excellent critiques of various drafts over several years. I am extremely grateful, in particular, to Pierre Dèbes for taking the time to give me such expert advice and judgment on so many topics. I am honored to be part of this community of mathematicians.

It has been a pleasure to work with Roger Astley and his staff at Cambridge University Press, and I thank them for their sound and professional counsel.

Finally, I acknowledge with gratitude the combined support of Davidson College, the Associated Colleges of the South, and the National Science Foundation.

Davidson, North Carolina
March 2004

Introduction

How to understand the numbers we encountered in secondary school, and equations involving them: this is our point of departure in studying Galois theory.

No two people have identical experiences in secondary school, to be sure; I would venture, however, that we all encountered numbers such as $1/7$, $\sqrt{2}$, $\sqrt[3]{-5}$, $\sqrt[4]{20}$, and $11 + 13/\sqrt{17}$. Now to begin a proper mathematical study of these numbers, we should consider what these numbers have in common – and which numbers we should exclude from our study. After all, a mathematical discipline proceeds by studying a little bit of mathematical reality quite closely, widening the field of vision only later.

A moment's reflection reveals that each of our numbers bears a certain relationship to rational numbers. Each is either a rational number, a root of a rational number, or some combination – using addition, subtraction, multiplication, and division – of rational numbers and roots of integral degree. Having made this observation, we might choose to take the plunge and restrict ourselves to arithmetic combinations of rational numbers and their roots, a set which would appear easy to manipulate.

Before rushing headlong into definitions and theorems, however, we should step back and contemplate whether we are comfortable with what it is that we are representing by the symbols above. For instance, what exactly do we mean by the symbol $\sqrt[3]{-5}$? *A priori*, all that we know of the number is that its cube is -5. An excellent question to ask at this point is whether or not such a number actually exists, and any answer to this question will depend, in some measure, on what we mean by the word *number*.

For the moment, let us simply ask whether or not there is, at least, some *complex number* such that its cube is -5. Our answer then is yes because, by the Fundamental Theorem of Algebra, inside the complex numbers exist roots of every polynomial (in one variable) with complex number coefficients. Hence there exists a complex number which is a root of $X^3 + 5$. Stated another way, there must be a complex number that is a solution to the equation $X^3 = -5$. We may agree, therefore, that when we think of a number, we will think of an element of the complex numbers.

We are not done, however, exploring what we mean by $\sqrt[3]{-5}$. After all, when we write $\sqrt[3]{-5}$ we are expressing only that we mean *some* root of the polynomial $X^3 + 5$, and there may exist several solutions – three, in fact. Our symbol $\sqrt[3]{-5}$, therefore, does not uniquely define a number. With this observation we face one of the dangerous subtleties in the naming of things.

To address this ambiguity, we now make a pact that when we write down a symbol for a number, we agree to specify that number as precisely as we can. Since there are three third roots of -5, we should provide another distinguishing characteristic of the number to indicate which of the three we mean. One distinguishing characteristic, for instance, is a complex approximation to the number. Only at the very end of the book, in section 35, will this pact expire, and adventurers there will have to decide amid the sound and fury of a grand generalization whether, in fact, what we signify there with our new definitions is nothing – or, somehow, everything.

Returning to our consideration of the numbers of secondary school, observe that we have isolated an important property of these numbers: they are not only complex numbers but also solutions to polynomial equations. It turns out that to think of rational numbers and their roots as part of a larger system of roots of polynomials is to give our work a more natural context. (We will return specifically to rational numbers and their roots in section 34, where we discuss solvability by radicals.)

Now we might choose to study the full set of numbers that are roots of polynomials, say polynomials with any complex coefficients whatsoever. Such a system, however, would cast the net extremely far out, since any complex number would be such a number. After all, if c is a complex number, it is certainly a root of the polynomial $X - c$. While the study of the arithmetic of the entire set of complex numbers is certainly compelling, we would quickly be caught short by the fact that there are complex numbers that we grasp very differently from those in our initial list.

Notice that, apart from rational numbers, we are able to express most complex numbers only by their *properties*. Furthermore, the nature of these properties typically dictates the way in which we study them. Even leaving aside the question of existence for numbers defined only by properties, we surely do not grasp such numbers or their "values" in the same sense as we grasp rational numbers, and the properties that complex numbers have may be quite varied.

For instance, we are familiar with the idea that i is a certain solution to the polynomial equation $X^2 = -1$, while π, on the other hand, is the ratio of the circumference of a circle to its diameter. It takes some work to associate a nongeometric property with π, such as, for instance, to see π as an infinite sum. Now to understand numbers defined by properties, we must look for ways to understand the connections between their properties. We have an enormous advantage with i, it turns out, since i is a root of polynomial with rational coefficients, and the fact that π is *not* the root of a polynomial with rational coefficients – in other words, the fact that π is *transcendental* – means that the methods of studying i are very likely not going to be especially useful in studying π.

In approaching Galois theory, we choose, then, to consider only those numbers that are roots of polynomials with *rational number* coefficients. Each of the numbers suggested at the beginning of this section satisfies this stronger criterion: 1/7 is a root of $7X - 1$; $\sqrt{2}$ is a root of $X^2 - 2$; $\sqrt[3]{-5}$ is a root of $X^3 + 5$; $\sqrt[4]{20}$ is a root of $X^4 - 20$, and $11 + 13/\sqrt{17}$ is a root of $X^2 - 22X + (1888/17)$. We call a root of a polynomial with rational coefficients an *algebraic number*.

Now that we have settled on a precise context for the numbers we wish to study, a context that is neither too narrow nor too broad, we turn to determining which equations involving algebraic numbers are valid. Immediately we ask whether one algebraic number may be expressed in terms of another. For instance, if ω is a nonreal third root of 1 – that is, a nonreal solution of $X^3 - 1 = 0$ – then we observe with interest that the other nonreal third root is ω^2, and, even further, that the three third roots are arithmetically related: $1 + \omega + \omega^2 = 0$. These observations cause us to wonder if there might be a *reduced form* of an expression involving algebraic numbers, so that by finding a unique reduced form we might decide if two sides of a purported equation are in fact equal. For instance, if we could reduce $2 + \omega^3$ and $4 + \omega + \omega^2$ to reduced forms, we might then notice that each is equal to 3.

These same observations will later lead us to ask whether this coincidence – that an expression involving one root of a polynomial is equal to another root of the same polynomial – is frequent or rare. Along the way we will consider the set of all expressions involving a particular root of a polynomial, calling this set a *field extension*, and we will wonder if the field extensions determined by two different roots of the same polynomial are somehow similar. Perhaps, under the right additional hypotheses, they are even isomorphic. In answering these questions, we will appreciate a group, the group of automorphisms of a field extension, that has been visible for only the past two centuries. The answers will also embrace an elegant correspondence between subsets of algebraic numbers and subgroups of Galois groups, a correspondence used to great effect by mathematicians today.

This text tells what is really only the first episode in the story of the algebraic numbers. We will review in the first chapter some preliminaries, and in the second chapter we will begin a close study of algebraic numbers. Moving into the third chapter, we will question what relationships exist among the many algebraic numbers, the polynomials of which they are roots, and the field extensions that they generate. The fourth chapter will show you how to consider more than one algebraic number at the same time, developing quite a bit of theory about isomorphisms, and then the fifth chapter will reveal the Galois correspondence. Along the way, pay particular attention to exercises marked with an asterisk, for they are referred to in the text, either beforehand or afterwards. Finally, for the adventurous who seek mathematical applications of the glorious correspondence, we offer several classical topics in the last chapter. Enjoy!

CHAPTER ONE

Preliminaries

This chapter briefly reviews some of the basic results and notation from a first course in abstract algebra that we need in our exposition of algebraic numbers and Galois theory. We also introduce a few functions from *Maple* and *Mathematica* that may assist the reader in exploring some of the material.

In this text, $\mathbb{N}$ denotes the integers greater than 0, and, given a field K, K^* denotes the multiplicative group of nonzero elements of K.

1. Polynomials, Polynomial Rings, Factorization, and Roots in $\mathbb{C}$

Definition 1.1 (Polynomial, Polynomial Ring). Let K be a field. The *polynomial ring* $K[X]$ *over* K is the set of formal sums

$$\left\{ \sum_{i=0}^{n} a_i X^i \ \bigg| \ a_i \in K, \ n \in \mathbb{N} \cup \{0\}, \ a_n \neq 0 \right\} \cup \{0\}.$$

Elements of $K[X]$ are called *polynomials over K*. Under the usual polynomial addition and multiplication, $K[X]$ is a commutative ring. The polynomial 0 is the additive identity, and the polynomial 1 is the multiplicative identity.

We usually denote polynomials by letters, but when we wish to indicate the underlying variable, we parenthesize the variable and append the expression to the name, as in $p(X)$.

A useful notion of the size of a nonzero polynomial over a field K is its degree.

Definition 1.2 (Degree of a Polynomial). Let K be a field and $p = p(X) = \sum_{i=0}^{n} a_i X^i$ a nonzero polynomial with $a_n \neq 0$. The *degree* $\deg(p)$ is n, the greatest power of X with nonzero coefficient in p.

The degree is therefore a function

$$\deg\colon K[X] \setminus \{0\} \to \mathbb{N} \cup \{0\} = \{0, 1, 2, \dots\}$$

satisfying $\deg(f + g) \leq \max\{\deg(f), \deg(g)\}$ and $\deg(fg) = \deg(f) + \deg(g)$ for $f, g \in K[X]$.

The degree of a polynomial p is 0 if and only if it is a nonzero element of K; hence $\deg(p) = 0 \Leftrightarrow p \in K^* \subset K[X]$. We call such polynomials, together with the polynomial 0, *constants*.

The analogy between polynomials and integers is one of the most fruitful in algebra, and in the following definitions and propositions we proceed to develop this analogy.

Definition 1.3 (Polynomial Factor, Reducible Polynomial). Let K be a field and $p \in K[X]$ a nonconstant polynomial. We say that p *factors over* K, or is *reducible over* K, if $p = fg$ for nonconstant polynomials $f, g \in K[X]$. Otherwise, p is *irreducible over* K.

We may omit the indication "over K" if the context makes its mention redundant. Note that we are uninterested in the case in which $p = fg$ with f or g an element of K since every $p \in K[X]$ may be so expressed: $p = (1/k)(kp)$ for any $k \in K^*$. We may multiply a nonzero polynomial p by an element of K in order to "normalize" it by changing the coefficient of its highest-order term to 1, just as for any nonzero integer we may always choose an element of $\{+1, -1\}$ by which to multiply the integer in order that the result is positive.

Definition 1.4 (Monic, Leading Coefficient). Let K be a field. A nonzero polynomial

$$0 \neq p = p(X) = \sum_{i=0}^{n} a_i X^i \in K[X]$$

is *monic* if its *leading coefficient* a_n is 1.

As with integers, we may divide one polynomial by another to produce a unique quotient and remainder.

Theorem 1.5 (Division Algorithm). *Let K be a field and $f, g \in K[X]$ polynomials with $f \neq 0$. Then we may constructively divide f into g so that there exist a unique quotient polynomial $q \in K[X]$ and a unique remainder polynomial $r \in K[X]$ such that*

- *$g = qf + r$ and*
- *either $\deg r < \deg f$ or $r = 0$.*

Proof. The algorithm follows by analogy the standard procedure for long division of integers, where in place of a decomposition of an integer into a sum of powers of 10, with coefficients ranging from 0 to 9, we decompose the polynomial into a sum of powers of X, with coefficients in K.

First we give a procedure that produces a q and r in $K[X]$ satisfying $g = qf + r$. If $g = 0$, then let $q = 0$ and $r = 0$. Otherwise, suppose

$$f = \sum_{i=0}^{\deg f} f_i X^i, \quad f_i \in K, \qquad g = \sum_{i=0}^{\deg g} g_i X^i, \quad g_i \in K.$$

If $\deg f > \deg g$, then let $q = 0$ and $r = g$, and we are done. Otherwise, we will find

$$q = \sum_{i=0}^{\deg(g) - \deg(f)} q_i X^i$$

with the $q_i \in K$ determined, one at a time, as follows.

Let $n = \deg(g) - \deg(f)$, and set q_n, the highest-order coefficient of q, to be the quotient of the highest-order coefficients of g and f, so that

$$q_n = g_{\deg g} / f_{\deg f}.$$

Then the polynomials g and $(q_n X^n) f$ agree in highest-order terms, and hence their difference,

$$d_n = g - (q_n X^n) f,$$

has degree no greater than $\deg(g) - 1$. If $n = 0$, then $\deg d_n < \deg f$ and we may stop after setting $r = d_n$.

Otherwise, we begin an induction on the coefficients of q. At each step, we define the coefficient q_{n-i} in such a way that $g - (q_n X^n + \cdots + q_{n-i} X^{n-i}) f$ has degree at most $\deg(g) - (i + 1)$. Clearly we have established the base case $i = 0$. Now assume that the induction is true for $i < n$.

Write d_{n-i} as

$$d_{n-i} = \sum_{j=0}^{\deg(g)-(i+1)} d_{n-i,j} X^j, \qquad d_{n-i,j} \in K$$

and set $q_{n-(i+1)}$ to be the quotient of certain coefficients of d_{n-i} and f:

$$q_{n-(i+1)} = d_{n-i,\deg(g)-(i+1)} / f_{\deg f}.$$

One checks that g and $(q_n X^n + \cdots + q_{n-(i+1)} X^{n-(i+1)}) f$ have identical coefficients for the terms with $X^{\deg(g)}, X^{\deg(g)-1}, \ldots, X^{\deg(g)-(i+1)}$. As a result, the difference

$$d_{n-(i+1)} = g - \left(\sum_{j=n-(i+1)}^{n} q_j X^j \right) f$$

has degree no greater than $\deg(g) - (i+2)$. Hence we have shown that the inductive statement is true for $i+1$. By the principle of mathematical induction, it is true for all $0 \le i \le n$ and we have defined a polynomial q.

By the induction property, $g - qf$ has degree no greater than $\deg(g) - (n+1) = \deg(f) - 1$. Letting $r = g - qf$, then, we have found a pair of polynomials q and r that satisfy the conclusions of the theorem.

Now we show that the q and r we constructed are unique. Suppose that there exist two pairs $q, r \in K[X]$ and $q', r' \in K[X]$ with

$$qf + r = g = q'f + r'$$

and each of r, r' is either zero or of degree less than $\deg f$. Then, subtracting the two representations of g, we have that the zero polynomial is equal to $(q - q')f + (r - r')$, or that

$$(q - q')f = r' - r.$$

If $(q - q')f$ is not the zero polynomial, then its degree is at least $\deg f$; however, if $r' - r$ is not zero, the degree of $r' - r$ is less than $\deg f$. Hence, if equality in $(q - q')f = r' - r$ is to hold, both sides must be the zero polynomial, which implies that $r = r'$ and $q = q'$.   □

Replacing the field K in the Division Algorithm with a larger field L (but keeping the same polynomials $f, g \in K[X] \subset L[X]$) *does not change* the outcome of the algorithm. However, the general question of whether or not a polynomial $f \in K[X]$ is reducible *does depend* on the field $L \supset K$: if L is sufficiently large, a polynomial irreducible over K may become reducible over L. For example, the polynomial $X^2 + 1$ is irreducible over $K = \mathbb{Q}$, but over a field L containing i (for instance, $L = \mathbb{C}$), $X^2 + 1$ factors into $X + i$ and $X - i$.

Just as with integers, we may define a greatest common divisor of two polynomials in $K[X]$ and find this greatest common divisor by means of a Euclidean Algorithm.

Definition 1.6 (Greatest Common Divisor I). Let K be a field and $f, g \in K[X]$ nonzero polynomials. A nonzero monic polynomial $p \in K[X]$ is the *greatest common divisor* $\gcd(f, g)$, or *GCD*, of f and g if p is a factor of both f and g, and, moreover, whenever a polynomial $h \in K[X]$ is a factor of both f and g, then h is a factor of p.

Theorem 1.7 (Euclidean Algorithm). *Let K be a field and $f, g \in K[X]$ nonzero polynomials. Then the greatest common divisor $\gcd(f, g) \in K[X]$ of f and g is the result of the following* Euclidean Algorithm.

Let $r_0 = f$ and $r_1 = g \in K[X]$, and set $i = 0$. Apply the Division Algorithm (Theorem 1.5) repeatedly for successively greater i to find $q_{i+2}, r_{i+2} \in K[X]$ such that $r_i = r_{i+1} q_{i+2} + r_{i+2}$, where $\deg r_{i+2} < \deg r_{i+1}$, until $r_{i+2} = 0$. Let j be the first index such that $r_j = 0$.

Then if a is the leading coefficient of r_{j-1}, then $(1/a) r_{j-1}$ is the greatest common divisor $\gcd(f, g)$ of f and g.

Working backwards, one may constructively express $\gcd(f, g)$ as a $K[X]$-linear combination of f and g, i.e., there constructively exist $z, w \in K[X]$ such that $\gcd(f, g) = zf + wg$.

Proof. It is an exercise (5.9) to show that the algorithm must terminate. We show first that r_{j-1} is a common divisor of f and g, and then we show that every common divisor of f and g divides r_{j-1}. Adjusting the coefficient a of the highest-order term, we find that $(1/a) r_{j-1}$ is then a monic polynomial that is the greatest common divisor of f and g.

From the last equation,

$$r_{j-2} = r_{j-1} q_j + r_j = r_{j-1} q_j,$$

we have that r_{j-1} divides r_{j-2}. Since each r_k, $0 \le k \le j - 2$, is defined to be a combination of r_{k+1} and r_{k+2}, it follows by induction that r_{j-1} divides every r_k, $0 \le k \le j - 2$. But then r_{j-1} divides $r_0 = f$ and $r_1 = g$. Hence r_{j-1} is a common divisor of f and g.

Going the other direction, suppose that a polynomial $h \in K[X]$ is a divisor of f and g. Then h divides $r_0 = f$ and $r_1 = g$. Since each r_k, $2 \le k \le j - 1$, is the remainder upon dividing r_{k-2} by r_{k-1}, it follows by induction that h divides every r_k, $0 \leq k \leq j-1$. But then h divides r_{j-1}.

It is an exercise (5.10) to show that $\gcd(f, g)$ may be expressed as a combination of f and g. □

It is an exercise (5.4) to prove that replacing K by a larger field L in the Euclidean Algorithm does not change its outcome.

Just as integers factor uniquely, up to a reordering of the factors, into a product of ± 1 and a set of primes, polynomials similarly factor in a unique way.

Theorem 1.8 ($K[X]$ is a Unique Factorization Domain). *Let K be a field. Then $K[X]$ is a unique factorization domain. In other words, every nonzero element $f \in K[X]$ has a factorization*

$$f = k \prod_{i=1}^{n} f_i, \qquad k \in K^*, \ 0 \neq f_i \in K[X],$$

where for each i, $\deg(f_i) \geq 1$ and f_i is monic and irreducible. Moreover, any such factorization of f is unique up to a reordering of the factors.

A proof of Theorem 1.8 based on the Euclidean Algorithm is an exercise (5.11).

The definition of a unique factorization domain is usually expressed more generally in terms of associates and irreducibles. Recall that an *integral domain* is a commutative ring with unity having no zero-divisors.

Definition 1.9 (Unique Factorization Domain). Let D be an integral domain. We say that $d \in D$ with $d \neq 0$ is *irreducible* if d is not a unit (i.e., is not invertible) and if $d = ab$ for $a, b \in D$, then either a or b is a unit. Two elements a, b of D are called *associates* if $a = ub$ for u a unit of D. We say that D is a *unique factorization domain* if (a) every nonzero element of D may be expressed as a product of irreducibles in D and (b) for each $d \in D$, all factorizations of d are equivalent by allowing permutation of the elements in the factorization and replacement of irreducibles by associates.

Knowing Theorem 1.8, we may define the greatest common divisor in an alternate fashion.

Definition 1.10 (Greatest Common Divisor II). Let K be a field and $f, g \in K[X]$ nonzero polynomials with factorizations

$$f = k_f \prod_{i=1}^{n_f} f_i, \qquad k_f \in K^*, \ 0 \neq f_i \in K[X],$$

$$g = k_g \prod_{i=1}^{n_g} g_i, \qquad k_g \in K^*, \ 0 \neq g_i \in K[X],$$

where each f_i and g_i is monic and irreducible. Let $P = \{f_i\} \cup \{g_i\}$ be the set of all irreducible factors that occur in either f or g, and for $p \in P$, let $\mathrm{ord}_f(p)$ denote the number of times p appears in the factorization of f, and likewise $\mathrm{ord}_g(p)$ the number of times p appears in the factorization of g. Note that $\mathrm{ord}_f(p)$ or $\mathrm{ord}_g(p)$ may be zero. The *greatest common divisor* $\gcd(f, g)$ is then

$$\gcd(f, g) = \prod_{p \in P} p^{\min(\mathrm{ord}_f(p),\ \mathrm{ord}_g(p))}.$$

Finally, just as each ideal of the integers $\mathbb{Z}$ consists of all integral multiples of some integer m, so does each ideal of $K[X]$ consist of all polynomial multiples of some polynomial m. (Recall that an *ideal* of a commutative ring is a subring closed under the operation of multiplying elements of the subring with elements of the ring.)

Definition 1.11 (Principal Ideal Domain). A *principal ideal domain* is an integral domain D in which every ideal I is *principal*, that is, for each ideal I there exists an element m of I such that $I = (m) = \{mf \mid f \in D\}$.

Theorem 1.12 ($K[X]$ is a Principal Ideal Domain). *Let K be a field. Then $K[X]$ is a principal ideal domain. Moreover, for each nonzero ideal I, $I = (m)$ for every nonzero m of minimal degree in I.*

Proof. Let $m \in I$ be a nonzero polynomial in I of minimal degree. Clearly $(m) \subset I$. Now let $g \in I$; we show that $g = mf$ for some $f \in K[X]$ as follows. First, use the Division Algorithm to write $g = fm + r$ for polynomials f and r in $K[X]$. Now $r = g - fm \in I$ and r is either the zero polynomial or of degree smaller than that of m; hence r must be the zero polynomial and $g \in (m)$. Therefore $I = (m)$. $\qquad \square$

Under the analogy we have developed, polynomials factor into monic irreducible factors just as an integer factors into primes. When the field K is sufficiently large, all irreducible factors of polynomials in $K[X]$ are of degree 1 (called *linear*), and in this case the analogy may be extended, identifying degree 1 factors of a polynomial with the corresponding roots. In this way a monic polynomial may be broken down into a set of roots with multiplicities, just as a positive integer may be broken down into a set of prime factors with multiplicities. If $K \subset \mathbb{C}$, then $\mathbb{C}$ is sufficiently large; this is the content of the Fundamental Theorem of Algebra.

Theorem 1.13 (Fundamental Theorem of Algebra). *Let $f \in \mathbb{C}[X]$ be a polynomial of degree at least 1. Then there exists a complex number $c \in \mathbb{C}$ such that $f(c) = 0$.*

The proof is beyond the scope of this text. However, we have the following:

Corollary 1.14. *Let $f \in \mathbb{C}[X]$ be irreducible. Then* $\deg(f) = 1$.

Proof of Corollary. Suppose that $\deg(f) > 1$. Then Theorem 1.13 gives us $c \in \mathbb{C}$ such that $f(c) = 0$. Let $g(X) = X - c$, and apply the Division Algorithm to write $f = qg + r$ with r a constant polynomial. Substituting c for X, we have that $f(c) = q(c)g(c) + r(c)$, which implies that $r(c) = 0$, so that r is the constant polynomial 0. Therefore $f = qg$ with q and g of degree at least one and therefore nonconstant, and hence f is reducible. $\square$

2. Computation with Roots and Factorizations: *Maple* and *Mathematica*

2.1. Approximating Roots

By repeatedly applying the Fundamental Theorem of Algebra and the Division Algorithm with polynomials of the form $X - c$, we deduce that every monic polynomial $f \in \mathbb{C}[X]$ may be written as a product of n linear factors $(X - c_i)$, where $c_i \in \mathbb{C}$ and $n = \deg(f)$. These c_i are the roots of f in $\mathbb{C}$. To calculate the numeric approximations to the set of roots of a polynomial in $\mathbb{C}[X]$, we use the command `fsolve(f=0,x,complex)` in *Maple* and `NSolve[f==0,x]` in *Mathematica*.

We will learn later that if the polynomial is irreducible over a subfield K of $\mathbb{C}$, then these roots c_i are all distinct. In that case, one may assign in some fashion an index from 1 to n to each root. In *Maple*, the `RootOf(f,index=n)` object represents the nth root of polynomial f, and we may approximate this root by supplying the object as an argument to the `evalf` function. In *Mathematica*, the `Root[f,n]` object represents the nth root of polynomial f, and we may approximate this root by supplying the object as an argument to the N function. (Note that *Maple* and *Mathematica* do not number the roots of a polynomial in the same fashion.)

We compute an approximation to the "first" and "second" roots of the polynomial $X^2 + 3$.

```
>   evalf(RootOf(x^2+3,index=1));
```
$$1.732050808\, I$$

```
In[1]:= N[Root[x^2 + 3, 1]]
```
$$Out[1] = 0. - 1.73205\, I$$

```
>   evalf(RootOf(x^2+3,index=2));
```
$$-1.732050808\, I$$

```
In[2]:= N[Root[x^2 + 3, 2]]
```
$$Out[2] = 0. + 1.73205\, I$$

We may alternatively find approximations to all of the roots without discovering *Maple*'s or *Mathematica*'s particular numbering of the roots.

```
>   fsolve(x^2+3=0,x,complex);
```
$$-1.732050808\, I,\ 1.732050808\, I$$

```
In[3]:= NSolve[x^2 + 3 == 0, x]
```
$$Out[3] = \{\{x \to -1.73205\, I\}, \{x \to 1.73205\, I\}\}$$

2.2. Factoring Polynomials over $\mathbb{Q}$

Irreducible polynomials play a prominent role in Galois theory, and it will be important for us to determine irreducibility. In *Maple*, the command to factor a polynomial over the rational numbers $\mathbb{Q}$ is `factor`, while in *Mathematica* the command is `Factor`.

```
>   factor(x^5+x+1);
```
$$(x^2 + x + 1)(x^3 - x^2 + 1)$$

```
In[4]:= Factor[x^5 + x + 1]
```
$$Out[4] = (1 + x + x^2)(1 - x^2 + x^3)$$

Factorization over $\mathbb{Q}$ poses quite a difficult computational problem. Irreducibility of polynomials can be verified by determining if the polynomial, with coefficients viewed modulo a prime p, is irreducible (see Exercise 5.16). When the polynomial is not irreducible, the standard route proceeds by taking the factorization of the polynomial modulo a prime p and using this information to find a factor of the polynomial modulo successively higher powers p^n of the prime. Finally, these successive approximations to the actual factor are used to determine a polynomial factor with integral coefficients. See the works on factorization in the bibliography for more information.

2.3. Executing the Division Algorithm over $\mathbb{Q}$

Another basic operation is that of dividing one polynomial in $\mathbb{Q}[X]$ into another to find the quotient and remainder polynomials. In *Maple*, the functions `quo` and `rem` find the quotient and remainder polynomials. In *Mathematica*, we use the functions `PolynomialQuotient` and `PolynomialRemainder` to find the quotient and remainder polynomials. These functions are valid over a variety of fields, including $\mathbb{Q}$, $\mathbb{R}$, and $\mathbb{C}$.

```
>   quo(2*x^3+3*x+1,x^2-1,x);
```
$$2\,x$$
```
>   rem(2*x^3+3*x+1,x^2-1,x);
```
$$1 + 5\,x$$

```
In[5]:= PolynomialQuotient[2x^3 + 3x + 1,
                x^2 − 1, x]

Out[5]= 2 x

In[6]:= PolynomialRemainder[2x^3 + 3x + 1, x^2 − 1, x]

Out[6]= 1 + 5 x
```

2.4. Executing the Euclidean Algorithm over $\mathbb{Q}$

Finding greatest common divisors with the Euclidean Algorithm will also be important to our work.

In *Maple*, the function is `gcd`. To determine additionally the polynomials z and w with which the greatest common divisor of f and g may be expressed in a linear combination $zf + wg$, we use the extended greatest common divisor function `gcdex`, providing additionally the variable name as well as two names in single quotation marks ("unevaluated" names) into which *Maple* will store the polynomials z and w. In the example below, we use `s` and `t` to store these and then check that the linear combination is in fact equal to the greatest common divisor.

In *Mathematica*, the function is `PolynomialGCD`. To determine additionally the polynomials z and w with which the greatest common divisor of f and g may be expressed in a linear combination $zf + wg$, we use the function `PolynomialExtendedGCD`, which requires the `PolynomialExtendedGCD` package. To load that package, execute `<<Algebra`PolynomialExtendedGCD`` first. The function returns both the greatest common divisor and a list of the polynomials z and w. Below we compute these coefficients and check that the linear combination is in fact equal to the greatest common divisor.

```
>   gcd(x^3-x^2+x-1,x^5-x^4+x^3-x^2+x-1);
```
$$x - 1$$
```
>   gcdex(x^3-x^2+x-1,x^5-x^4+x^3-x^2+x-1,x,
>       's','t');
```
$$-1 + x$$
```
>   s; t;
```
$$-x^2$$
$$1$$
```
>   expand(s*(x^3-x^2+x-1)+
>       t*(x^5-x^4+x^3-x^2+x-1));
```
$$-1 + x$$

In[7]:= PolynomialGCD[x^3 − x^2 + x − 1,

x^5 − x^4 + x^3 − x^2 + x − 1]

Out[7]= −1 + x

In[8]:= << Algebra`PolynomialExtendedGCD`

In[9]:= PolynomialExtendedGCD[

x^3 − x^2 + x − 1,

x^5 − x^4 + x^3 − x^2 + x − 1]

Out[9]= {−1 + x, {−x², 1}}

In[10]:= Expand[(−x^2) * (x^3 − x^2 + x − 1)+

1 * (x^5 − x^4 + x^3 − x^2 + x − 1)]

Out[10]= −1 + x

3. Ring Homomorphisms, Fields, Monomorphisms, and Automorphisms

The basic property of homomorphisms of rings that we will use is the First Isomorphism Theorem.

Theorem 3.1 (First Isomorphism Theorem for Rings). *Let $\varphi: R \to S$ be a homomorphism of rings. Then the function*

$$\bar{\varphi}: R/\ker\varphi \to \varphi(R)$$

given by

$$\bar{\varphi}(r + \ker\varphi) = \varphi(r)$$

is a ring isomorphism.

Proof. First we check that $\bar{\varphi}$ is well defined, that is, the function does not depend on the representative r of the coset $r + \ker\varphi$. If $r + \ker\varphi = r' + \ker\varphi$, then by properties of cosets, $r - r' \in \ker\varphi$. But then

$$\bar{\varphi}(r + \ker\varphi) - \bar{\varphi}(r' + \ker\varphi) = \varphi(r) - \varphi(r') = \varphi(r - r') = 0.$$

Hence $\bar{\varphi}$ is well defined. Checking that $\bar{\varphi}$ is a homomorphism is routine.

Now a similar argument shows that $\bar{\varphi}$ is one-to-one, as follows. If $\bar{\varphi}(r + \ker\varphi) = \bar{\varphi}(r' + \ker\varphi)$, then $\varphi(r) = \varphi(r')$ and $\varphi(r - r') = 0$. Hence $r - r' \in \ker\varphi$ and the cosets $r + \ker\varphi$ and $r' + \ker\varphi$ are identical. Finally, $\bar{\varphi}$ clearly maps $R/\ker\varphi$ onto $\varphi(R)$. Hence $\bar{\varphi}$ is an isomorphism.   □

We will frequently consider the case when the rings in question are fields. Now the only ideals of a field F are $\{0\}$ and F itself, and consequently the kernel of any ring homomorphism $\varphi\colon F \to S$ must be either $\{0\}$ or F. In the latter case, φ clearly sends every element to 0, and we call such a homomorphism *trivial*. Otherwise, $\ker\varphi$ is $\{0\}$ and therefore the homomorphism is one-to-one. For this reason, we usually speak of field *monomorphisms* (i.e., one-to-one homomorphisms) instead of field homomorphisms.[1] Isomorphisms from a field F to itself are called *automorphisms*, and the set of all automorphisms of F is denoted $\mathrm{Aut}(F)$.

Sometimes we will want to apply a field monomorphism $\varphi\colon F \to F'$ between fields F and F' to the coefficients of a polynomial in $F[X]$. Doing so is the same as applying a natural extension of the monomorphism to polynomial rings over the two fields, and it is routine to check that the map in the following definition is a monomorphism.

Definition 3.2 (Extension of Monomorphism to Polynomial Rings). Let F and F' be two fields and $\varphi\colon F \to F'$ a field monomorphism. Then the map

$$\varphi\left(a_n X^n + a_{n-1} X^{n-1} + \cdots + a_0\right) = \varphi(a_n) X^n + \varphi(a_{n-1}) X^{n-1} + \cdots + \varphi(a_0)$$

is a monomorphism from $F[X]$ to $F'[X]$, which we also name φ.

---

[1] Onto homomorphisms are also called *epimorphisms*, but we will not use this term in this text.

4. Groups, Permutations, and Permutation Actions

Definition 4.1 (Generated Group). A subgroup H of G is *generated by* elements $\epsilon_1, \ldots, \epsilon_n$ if H is the smallest subgroup of G containing $\epsilon_1, \ldots, \epsilon_n$. We write that $H = \langle \epsilon_1, \ldots, \epsilon_n \rangle$.

Analogous definitions apply for generated subrings and generated ideals of ring, as well as for generated subfields of a field. We will mention these in greater detail as they arise.

Definition 4.2 (Permutation). Let S be a finite set. A *permutation* of S is a one-to-one and onto function from S to S.

Definition 4.3 (Symmetric Group). Let n be a positive integer. The *symmetric group* S_n is the set of all permutations of the set of n "letters," usually written $\{1, 2, \ldots, n\}$.

Definition 4.4 (Permutation Group). A *permutation group* is a group such that the underlying set consists of permutations of some finite set S. A *permutation representation* of a group G is an isomorphism between G and group of permutations of a finite set S, and often the group G is then considered to be identified with this group of permutations. When a group has a permutation representation on a finite set S, we say that G *acts on S*. When $|S| = n$, so that the set S may be identified with the set of n letters $\{1, 2, \ldots, n\}$, then G may be identified with a subgroup of S_n.

Definition 4.5 (Orbit, Stabilizer). Let G be a permutation group on S. The *orbit* $\mathrm{orb}_G(s)$ of an element $s \in S$ is the set

$$\mathrm{orb}_G(s) = \Big\{ g(s) \mid g \in G \Big\}.$$

The *stabilizer* $\mathrm{stab}_G(s)$ of an element $s \in S$ is the subgroup

$$\mathrm{stab}_G(s) = \Big\{ g \in G \mid g(s) = s \Big\}$$

of G.

Theorem 4.6 (Orbit-Stabilizer Theorem). *Let G be a permutation group on S, $s \in S$, and $H = \mathrm{stab}_G(s)$. Then $|\mathrm{orb}_G(s)| = |G|/|H|$.*

Proof. The group may be partitioned into $|G|/|H|$ cosets gH, $g \in G$. We claim that the action of the elements of gH on s is identical, sending s to $g(s)$, and that elements from different cosets send s to different images. This will prove the theorem.

Let $gh_1, gh_2 \in gH$ for a coset gH. Then $(gh_1)(s) = g(h_1(s)) = g(s)$ and $(gh_2)(s) = g(h_2(s)) = g(s)$, so our first claim is proved. For the second, suppose that $g_1 h_1 \in g_1 H$ and $g_2 h_2 \in g_2 H$ send s to the same element: $(g_1 h_1)(s) = (g_2 h_2)(s)$. Then $(g_1 h_1)(s) = g_1(h_1(s)) = g_1(s)$, and this is equal to $(g_2 h_2)(s) = g_2(h_2(s)) = g_2(s)$. But then $g_1^{-1} g_2$ leaves s fixed, so $g_1^{-1} g_2 \in H$, implying that the cosets $g_1 H$ and $g_2 H$ are equal. □

5. Exercises

Problems that are referred to in the text are denoted by an asterisk.

A.

5.1 Use the Division Algorithm to divide into quotients and remainders the following pairs of polynomials. Show the intermediate q_i and d_i at each step, and check your results with *Maple* or *Mathematica*.

(1) $X^2 - 4$ divided by $X - 2$ in $\mathbb{Q}[X]$;
(2) $X^3 - 1$ by $2X - 2$ in $\mathbb{Q}[X]$;
(3) $X^3 + X^2 + X + 1$ by $5X + 4$ in $\mathbb{Q}[X]$;
(4) X by $X - 1$ in $\mathbb{Q}[X]$;
(5) $X^2 - 2$ by $\sqrt{2}X - 2$ in $\mathbb{R}[X]$;
(6) $X^2 + 1$ by $X - i$ in $\mathbb{C}[X]$.

5.2 Use the Euclidean Algorithm to find the greatest common divisors of the following pairs of polynomials. Show the intermediate q_i and r_i at each step, and check your results with *Maple* or *Mathematica*.

(1) $X^2 - 4$ and $X^2 + 4X + 4$ in $\mathbb{Q}[X]$;
(2) $X^3 - 1$ and $2X^5 - 2$ in $\mathbb{Q}[X]$;
(3) $X^3 - 1$ and $X^2 + X + 1$ in $\mathbb{Q}[X]$;
(4) $X^2 + 2$ and $X^2 - 2$ in $\mathbb{R}[X]$;
(5) $X^3 - 2iX^2 + 7X + 4i$ and $X^3 - 3iX^2 + 7X + 3$ in $\mathbb{C}[X]$.

5.3 Determine which of the following polynomials are irreducible over $\mathbb{Q}$.

(1) $X^3 + 1$;

(2) $X^4 - X + 1$;

(3) $X^4 - X^2 + X - 1$;

(4) $X^5 + X + 1$.

5.4* Prove that replacing the field K by a field $L \supset K$ does not change the outcome of the Euclidean Algorithm, and hence the notion of the greatest common divisor is independent of the field K containing the coefficients of polynomials $f, g \in K[X]$.

5.5 Suppose $f, g, h \in K[X]$, f and g are nonzero, $\gcd(f, g) = 1$, and f and g both divide h. Show that fg divides h.

5.6 Prove that the following are equivalent:

- $p \in K[X]$ is irreducible;
- $(p(X))$ is a maximal ideal in $K[X]$;
- $K[X]/(p(X))$ is a field.

5.7* Prove that if $K \subset \mathbb{Q}$ is a field, then $K = \mathbb{Q}$. (For an arbitrary field K, the smallest subfield F of K is called its *prime subfield*, and F is the subfield generated by 1. If F is finite, then $|F| = p$ for a prime p and we say that K has *characteristic p*. If F is infinite, we say that K has *characteristic* 0.)

B.

5.8 Prove that if f is irreducible over K and $c \in K^*$, then $f(X + c)$, $f(cX)$, and $cf(X)$ are all irreducible over K.

5.9* Prove that the Euclidean Algorithm terminates.

5.10* Prove (by induction on the number of remainders in the Euclidean Algorithm, if desired) that the greatest common divisor of two polynomials $f, g \in K[X]$ may be expressed as a $K[X]$-linear combination $zf + wg$, with $z, w \in K[X]$, of f and g.

5.11* Prove Theorem 1.8 using the Euclidean Algorithm, as follows: (a) show (by induction on the degree, if desired) that any nonconstant polynomial in $K[X]$ is a product of irreducible polynomials; (b) show that if $f \in K[X]$ is irreducible and f divides gh, where $g, h \in K[X]$, then f divides g or f divides h, by proving that the greatest common divisor of f and g is either a constant multiple of f or the constant 1, expressing h as a combination of fh and gh in the second case; (c) use part (b) to show that two factorizations into irreducibles $f_1 f_2 \cdots f_n = g_1 g_2 \cdots g_m$ have the same irreducible components up to constants.

5.12 Prove that Definitions 1.6 and 1.10 are equivalent. (First show that the GCD is well defined by the properties in the first definition: precisely one polynomial satisfies the properties. Then show that this polynomial is in fact that given by the second definition.)

5.13 Prove that several definitions of the *least common multiple* lcm(f, g) of two polynomials $f, g \in K[X]$ over a field K are equivalent: first, the least common multiple lcm(f, g) of $f, g \in K[X]$ is the unique monic generator of the intersection of the ideals (f) and (g) of $K[X]$; second, the least common multiple lcm(f, g) of $f, g \in K[X]$ is the "normalized" polynomial for $fg/\gcd(f, g)$, that is, the polynomial $fg/\gcd(f, g)$ divided by its leading coefficient; and finally, an appropriate analogue of Definition 1.10.

5.14* Prove the *Division Algorithm for Integral Domains*: Let R be an integral domain and $f, g \in R[X]$ with f a nonzero polynomial having a unit in R for its leading coefficient. Then we may constructively divide f into g so that there exist a unique quotient polynomial $q(X) \in R[X]$ and a unique remainder polynomial $r(X) \in R[X]$ such that $g = qf + r$ and $\deg r < \deg f$ or $r = 0$. (Consider the proof of Theorem 1.5.)

5.15 Prove the *Rational Root Theorem*: If a polynomial with integer coefficients

$$f = a_n X^n + a_{n-1} X^{n-1} + \cdots + a_0 \in \mathbb{Z}[X]$$

has a root in $\mathbb{Q}$, then the root takes the form r/s, where r is a factor of a_0 and s is a factor of a_n.

C. Since we use *Maple* or *Mathematica* to factor polynomials, we will not generally need the following criteria. Nevertheless, they remain of high theoretical interest and are used in the service of proving a variety of results.

5.16* Let p be a prime, $\mathbb{F}_p = \mathbb{Z}/p\mathbb{Z}$ the finite field with p elements, $f \in \mathbb{Z}[X]$ a polynomial, and let $\varphi \colon \mathbb{Z}[X] \to \mathbb{F}_p[X]$ be defined by

$$\varphi\left(\sum_{i=0}^{n} a_i X^i\right) = \sum_{i=0}^{n} (a_i \bmod p)\, X^i.$$

Prove the *Mod p Irreducibility Criterion*: If $\varphi(f)$ is irreducible over $\mathbb{F}_p$ and $\deg(f) = \deg(\varphi(f))$, then f is not the product of two polynomials $g, h \in \mathbb{Z}[X]$ of degree at least 1.

It is a fact that if $f \in \mathbb{Z}[X]$ is reducible over $\mathbb{Q}$, then f is reducible over $\mathbb{Z}$. Hence if $\varphi(f)$ is irreducible over $\mathbb{F}_p$ and $\deg(f) = \deg(\varphi(f))$, then f is irreducible over $\mathbb{Q}$.

5.17 Prove the *Eisenstein Irreducibility Criterion*: Suppose that

$$f = a_n X^n + a_{n-1} X^{n-1} + \cdots + a_0 \in \mathbb{Z}[X]$$

is a polynomial with integer coefficients and that p is a prime such that

   (1) p divides a_i for all $i < n$;

   (2) p does not divide a_n;

   (3) p^2 does not divide a_0.

Then f is irreducible over $\mathbb{Q}$. (Hint: Work by contradiction with the following sketch. It is a fact that if $f \in \mathbb{Z}[X]$ is reducible over $\mathbb{Q}$, then f is reducible over $\mathbb{Z}$. We may therefore assume without loss of generality that if f factors into polynomials g and h in $\mathbb{Q}[X]$, with coefficients $\{b_i\}$ and $\{c_i\}$, respectively, then these coefficients are integers. Without loss of generality, the constant term of h is divisible by p. The leading coefficient of h cannot be divisible by p, so we may consider the coefficient of smallest degree of h not divisible by p, say of degree i. If $i < n$, then p divides a_i and then $b_0 c_i = a_i - (b_1 c_{i-1} + \cdots + b_i c_0)$ is divisible by p, which is a contradiction.)

CHAPTER TWO

Algebraic Numbers, Field Extensions, and Minimal Polynomials

In this chapter, we introduce algebraic numbers and two mathematical objects associated to algebraic numbers: field extensions and minimal polynomials. We develop these, especially the field generated by an algebraic number, as concretely as possible. As a result, the exposition is slow and deliberate. (If you are already familiar with some of the results in this chapter, you should be able to absorb it quickly and begin to work some of the exercises.) We conclude with a theorem connecting fields generated by algebraic numbers and the less concrete notion of quotients of polynomial rings.

6. The Property of Being Algebraic

Definition 6.1 (Algebraic Number). An *algebraic number* α is a complex number $\alpha \in \mathbb{C}$ such that there exists a polynomial $0 \neq p \in \mathbb{Q}[X]$ with $p(\alpha) = 0$. Such polynomials p are said to be *associated* to α.

As mentioned in the introduction, the set of algebraic numbers provides a good context in which to study certain numbers encountered in secondary school: rational numbers, such as $3/7$, and their integral roots, such as $\sqrt[2]{7}$ and $\sqrt[3]{14}$. These numbers are certainly roots of polynomials with rational coefficients, namely, $X - (3/7)$, $X^2 - 7$, and $X^3 - 14$. In fact, they are all roots of polynomials of a particularly simple form, $X^n - a$, where $a \in \mathbb{Q}$ and n is a positive integer. The sums and products of these numbers, however, are generally not roots of polynomials of this simple form, so to consider sums and products, we must expand our domain to include algebraic numbers as defined above. We will see later that the set of algebraic numbers is closed under addition and multiplication and is therefore a useful domain in which to work.

Algebraic numbers are not, in general, determined uniquely by their associated polynomials. For instance, two different algebraic numbers may share an associated polynomial, and, moreover, two algebraic numbers may share all of their associated polynomials. For example, there is no polynomial $p \in \mathbb{Q}[X]$ such that $p(\sqrt{2}) = 0$ but $p(-\sqrt{2}) \neq 0$. Hence we find in the associated polynomials of an algebraic number α some important properties, but by no means the only important properties, of α.

Before going on, we note that not all elements of $\mathbb{C}$ are algebraic. This was proved by Liouville in 1844 [55], and the fact that π, in particular, is not algebraic was proved by Lindemann in 1882 [54]. Nonalgebraic numbers are called *transcendental*. For details, see [21, §III.14] or [25, Chap. 24].

7. Minimal Polynomials

Each algebraic number α has many polynomials $p \in \mathbb{Q}[X]$ associated to it. Our goal in this section is to describe the set of polynomials associated to α and show that this set can be described in terms of a particular polynomial m_α associated to α, called its minimal polynomial.

Definition 7.1 (Minimal Polynomial, Degree of α). Let α be an algebraic number. The *minimal polynomial* m_α is the unique monic, irreducible polynomial in $\mathbb{Q}[X]$ having α as a root. If K is a field containing $\mathbb{Q}$, then the *minimal polynomial* $m_{\alpha,K}$ *over* K *of* α is the unique monic, irreducible polynomial in $K[X]$ having α as a root. We say that the *degree of* α is the degree $\deg(m_\alpha)$ and denote it by $\deg(\alpha)$; similarly, we say that the *degree of* α *over* K is the degree $\deg(m_{\alpha,K})$ and denote it by $\deg_K(\alpha)$.

If $K = \mathbb{Q}$ we have $m_{\alpha,\mathbb{Q}} = m_\alpha$ and $\deg_\mathbb{Q}(\alpha) = \deg(\alpha)$.

We now prove that Definition 7.1 is valid, showing that such unique monic, irreducible polynomials exist, and show moreover that the set of polynomials associated to an algebraic number α is determined by m_α.

Proposition 7.2 (Polynomials Associated to α over K). *Let K be a subfield of $\mathbb{C}$, and let α be an algebraic number.*

The set of polynomials in $K[X]$ associated to α, together with the zero polynomial, is an ideal $I_{\alpha,K}$ of $K[X]$, and in $I_{\alpha,K}$ there exists a unique monic, irreducible (over K) polynomial $m_{\alpha,K}$. The ideal is generated by $m_{\alpha,K}$; that is, every polynomial in the ideal is the product of $m_{\alpha,K}$ and a polynomial in $K[X]$.

We express these statements with the following notation:

$$I_{\alpha,K} = \left\{ p \in K[X] \mid p(\alpha) = 0 \right\} = (m_{\alpha,K}(X)) = \left\{ f m_{\alpha,K} \mid f \in K[X] \right\}.$$

Proof. A proof of the first fact, that the set $I_{\alpha,K}$ of polynomials associated to α, together with the zero polynomial, is an ideal, is an exercise (10.1).

We now show that $I_{\alpha,K}$ is generated by a monic, irreducible polynomial. Among the set of polynomials in $K[X]$ associated to α, there must exist some polynomial with smallest degree, say p. We claim that p is irreducible over K. Suppose not; then there exist nonconstant polynomials $g, h \in K[X]$ of smaller degree such that

$$p = gh.$$

Substituting α for X, the left-hand side must reduce to zero; hence $g(\alpha)h(\alpha) \in \mathbb{C}$ must also be zero, whence either g or h is a polynomial associated to α of degree smaller than $\deg p$, contrary to assumption. Therefore p is an irreducible polynomial associated to α. Let a_n denote the leading coefficient of p, and define $\bar{p} = (1/a_n)p$. Then $\bar{p}$ is necessarily monic and irreducible and of minimal degree in I. Theorem 1.12 tells us that each ideal I of $K[X]$ is generated by any polynomial of minimal degree in I; hence $I_{\alpha,K} = (\bar{p})$.

Now this monic, irreducible polynomial associated to α is unique, as follows. Suppose that $g \in K[X]$ is a monic, irreducible polynomial associated to α. Apply the Division Algorithm (Theorem 1.5) to divide $\bar{p}$ into g, resulting in $g = \bar{p}q + r$ for polynomials $q, r \in K[X]$ with $r = 0$ or $\deg r < \deg \bar{p}$. Now $r = g - \bar{p}q$ is an element of the ideal $I_{\alpha,K}$, and if $r \neq 0$, then by substituting α for X in the polynomials, we verify that r is a polynomial associated to α of degree smaller than $\deg \bar{p}$, contrary to assumption. Hence $g = \bar{p}q$ for nonzero polynomials $\bar{p}$ and q. Now since g is irreducible, either $\bar{p}$ or q is constant. If $\bar{p}$ is a nonzero constant, then p has no roots in $\mathbb{C}$, a contradiction; hence q is a constant. Since both $\bar{p}$ and g are monic, $q = 1$. Hence $g = \bar{p}$. Then $m_{\alpha,K} = \bar{p}$ is the unique monic, irreducible polynomial associated to α. $\square$

8. The Field Generated by an Algebraic Number

In order to understand an algebraic number α over a field K with $\mathbb{Q} \subset K \subset \mathbb{C}$, it is useful to understand the collection of all numbers that may be represented by some combination of α and elements of K using the four *field operations*: addition, subtraction, multiplication, and division. (Note that the interest is when $\alpha \notin K$.)

In the following two definitions, we distinguish between, on one hand, formal *expressions* that involve the field operations, some elements of K, and α, calling these *arithmetic combinations*; and, on the other, the elements of $\mathbb{C}$ that correspond to the *evaluation* of these formal expressions. We say that a number α is *represented* by an arithmetic combination if the arithmetic combination evaluates to α.

Definition 8.1 (Arithmetic Combination). Let K be a subfield of $\mathbb{C}$ and α an algebraic number. An *arithmetic combination of α over K* is either α, an element of K, or a formal sum, difference, product, or quotient of two arithmetic combinations, subject to the following two constraints: (a) in the formation of an arithmetic combination, we do not permit taking the quotient of an arithmetic combination and zero; and (b) we require that the total number of sums, differences, products, and quotients in an arithmetic combination be finite.

Note that α, $1 \cdot \alpha$, and $(\alpha + 2) - 2$ are examples of arithmetic combinations all representing the same number as α.

Definition 8.2 (Generated Field). Let K be a subfield of $\mathbb{C}$ and α an algebraic number. The *field $K(\alpha)$ generated by α over K* is the set of all numbers in $\mathbb{C}$ that are represented by arithmetic combinations of α over K. If $K = \mathbb{Q}$, then we call $\mathbb{Q}(\alpha)$ the *field generated by α*. We also say that $K(\alpha)$ is the field K with α *adjoined*.

In this section, we prove that $K(\alpha)$ is indeed a field. In doing so, we show how to construct a unique, reduced form for any element of $K(\alpha)$, which allows us to decide if two arithmetic combinations are equal. We will see later that when K contains only algebraic numbers, then every element of $K(\alpha)$ is an algebraic number.

The field $K(\alpha)$ will also be our first example of a *field extension*; under the following definition, $K(\alpha)$ is said to be a field extension of K.

Definition 8.3. Let K and L be two fields. If $K \subset L$, then we say that L is a *field extension* of K or that L is a field *over K*.

8.1. Rings and Vector Spaces Associated to an Algebraic Number

We first seek to understand the structure of the set of all numbers in $\mathbb{C}$ that may be represented by certain arithmetic combinations of α over K – combinations that, in their construction, contain only *three* of the four types of operations, namely, additions, subtractions, and multiplications. We will discover later that this set represents every number in the field $K(\alpha)$, even though *a priori* the field $K(\alpha)$ may be larger, since it also includes numbers represented by arithmetic combinations that contain quotients.

Definition 8.4 (Generated Ring). Let K be a subfield of $\mathbb{C}$ and α an algebraic number. The *ring $K[\alpha]$ generated by α over K* is the set of all numbers in $\mathbb{C}$ that may be represented by arithmetic combinations of α over K without taking quotients. If $K = \mathbb{Q}$, then we call $\mathbb{Q}[\alpha]$ the *ring generated by α*.

By analogy with the definition of a generated group (Definition 4.1), we may equivalently define the ring as the smallest subring of $\mathbb{C}$ containing α.

Example 8.5. Let $\alpha = \sqrt{2}$ and $K = \mathbb{Q}$. Then $\beta = \sqrt{2}/2$ lies in $K[\alpha]$ and hence also in $K(\alpha)$, since β may be expressed using multiplication of an element of K and α: $\beta = \frac{1}{2} \cdot \sqrt{2}$. Similarly, $\gamma = 1/\sqrt{2}$ certainly lies in $K(\alpha)$. One might expect γ not to lie in $K[\alpha]$, since $\sqrt{2}$ appears in a denominator. However, $\gamma = 1/\sqrt{2} = \sqrt{2}/2 = \beta$, so γ lies in $K[\alpha]$ as well.

Observe that each arithmetic combination of α over K without quotients may be expanded out to a "polynomial in α": an expression of the form $\sum_{i=0}^{n} c_i \alpha^i$ for some $n \geq 0$ and $c_i \in K$. (Exercise 10.20 suggests an algorithmic proof of this fact, based on the concept of the number of operations in the arithmetic combination.) Conversely, any such expression $\sum_{i=0}^{n} c_i \alpha^i$ is surely an arithmetic combination of α over K. We conclude that $K[\alpha]$ is the set of values at α of all the polynomials in $K[X]$.

That the set $K[\alpha]$ is a commutative ring is Exercise 10.10.

Theorem 8.6. *Let K be a subfield of $\mathbb{C}$ and α an algebraic number. Each element of $K[\alpha]$ has a unique representation as a sum of K-multiples of nonnegative powers of α, where the* *powers are less than* $\deg_K(\alpha)$:

$$K[\alpha] = \left\{ \sum_{i=0}^{\deg_K(\alpha)-1} k_i \alpha^i \ \middle| \ k_0, k_1, \ldots, k_{(\deg_K(\alpha)-1)} \in K \right\}. \tag{8.1}$$

This representation will be called the reduced form in α *for elements of* $K[\alpha]$. *Furthermore, under the standard operations of addition and multiplication of complex numbers,* $K[\alpha]$ *is a vector space over* K *of dimension* $\deg_K(\alpha)$:

$$[K[\alpha] : K] = \deg_K(\alpha).$$

Proof. By Exercise 10.20, each arithmetic combination of α over K representing a number $\beta \in \mathbb{C}$ may be replaced by another arithmetic combination, also in α over K and representing the same β, in polynomial form:

$$\beta = \sum_{i=0}^{n} c_i \alpha^i, \qquad n \geq 0, \ c_i \in K. \tag{8.2}$$

Substitute X for α in equation (8.2) to produce a polynomial $f = f(X) \in K[X]$. Apply the Division Algorithm (Theorem 1.5) to divide f by $m_{\alpha,K}(X)$, and let $q = q(X)$ be the quotient and $r = r(X)$ the remainder. The remainder r is either zero or of degree at most $\deg_K(\alpha) - 1$. Now substitute α for X in r. We then have an expression

$$r(\alpha) = \sum_{i=0}^{\deg_K(\alpha)-1} k_i \alpha^i, \qquad k_i \in K.$$

The expression $r(\alpha)$ is yet another arithmetic combination evaluating to β, as follows. By the Division Algorithm, $f(X) = q(X)m_{\alpha,K}(X) + r(X)$. Substituting α for X, we have $f(\alpha) = q(\alpha)m_{\alpha,K}(\alpha) + r(\alpha)$. Since $f(\alpha) = \beta$ and $m_{\alpha,K}(\alpha) = 0$ by definition of the minimal polynomial of α over K, we have that $\beta = r(\alpha)$. We have shown then that a number that is represented by an arithmetic combination not containing a quotient may be expressed in the form of the theorem.

Now we show that among all arithmetic combinations in α taking polynomial form and of degree at most $\deg_K(\alpha) - 1$, this combination is the unique one that represents β. Suppose that there exist two expressions for β

$$\beta = \sum_{i=0}^{\deg_K(\alpha)-1} k_i \alpha^i, \qquad k_i \in K \qquad \text{and} \qquad \beta = \sum_{i=0}^{\deg_K(\alpha)-1} l_i \alpha^i, \qquad l_i \in K.$$

Taking the difference of the two expressions, we have

$$0 = \sum_{i=0}^{\deg_K(\alpha)-1} (k_i - l_i)\alpha^i, \qquad (k_i - l_i) \in K. \tag{8.3}$$

Substituting X for α in the right-hand side of equation (8.3), however, we have a polynomial in $K[X]$ that lies in $I_{\alpha,K}$. This polynomial has degree less than $\deg_K(\alpha)$, which is the minimal degree of any nonzero polynomial in $I_{\alpha,K} = (m_{\alpha,K}(X))$ (Theorem 1.12); therefore the polynomial is the zero polynomial. But then $k_i - l_i = 0$ for each $i = 0, 1, \ldots, \deg_K(\alpha) - 1$. Hence $k_i = l_i$ and the two expressions are identical.

Finally, we claim that the ring $K[\alpha]$ is a vector space over K of dimension $\deg_K(\alpha)$. It is an exercise (10.11) to show that the axioms of a vector space are satisfied for $K[\alpha]$ over K. We determine its dimension by showing that

$$\left\{ 1, \alpha, \alpha^2, \ldots, \alpha^{\deg_K(\alpha)-1} \right\}$$

is a basis for $K[\alpha]$ over K. We have already shown that every element of $K[\alpha]$ is a linear combination of the elements of this set, with coefficients in K; hence the set spans $K[\alpha]$. To show linear independence, suppose that

$$0 = \sum_{i=0}^{\deg_K(\alpha)-1} k_i \alpha^i, \qquad k_i \in K,$$

for some k_i. Proceeding as in the latter portion of the last paragraph, we have that all k_i must be zero. Hence the set is linearly independent and so is a basis. The dimension over K is therefore $\deg_K(\alpha)$.   □

Now addition and subtraction, performed term-by-term with reduced forms $\sum_{i=0}^{\deg_K(\alpha)-1} c_i \alpha^i$, clearly result in reduced forms. Furthermore, the proof of the theorem suggests a method for obtaining unique expressions for numbers in $K[\alpha]$ even when multiplying: multiply the two expressions and expand into the polynomial form $\sum_{i=0}^{2\deg_K(\alpha)-2} c_i \alpha^i$, replace α temporarily by an indeterminate X, divide the result by $m_{\alpha,K}(X)$, and retain the remainder, replacing X by α. Preserving reduced expressions in this fashion through addition, subtraction, and multiplication will be termed *reduction modulo $m_{\alpha,K}$*.

Example 8.7. Let $m_{\alpha,\mathbb{Q}} = X^3 - 2$, or, equivalently,[1] write $\alpha = \sqrt[3]{2}$. Suppose that we wish to find the reduced form for

$$(\alpha^3 + \alpha)\alpha^2 + 2\alpha - (\alpha + 1).$$

We first expand the expression into a sum of K-multiples of nonnegative powers of α:

$$\alpha^5 + \alpha^3 + \alpha - 1.$$

We now view this expression as a polynomial in an "indeterminate" α, dividing it by $m_{\alpha,\mathbb{Q}}(\alpha)$, which gives us

$$\alpha^5 + \alpha^3 + \alpha - 1 = (\alpha^2 + 1)(\alpha^3 - 2) + (2\alpha^2 + \alpha + 1).$$

The remainder, $2\alpha^2 + \alpha + 1$, is then the reduced form, for it is a sum of K-multiples of nonnegative powers of α, where these powers do not exceed $\deg(\alpha) - 1 = 2$.

Remark 8.8. It is sometimes useful to view the polynomial $m_{\alpha,K}(X)$ as a *relation* between the monic highest term X^n and the succeeding terms: if

$$m_{\alpha,K}(X) = X^n + a_{n-1}X^{n-1} + \cdots + a_0,$$

then we have an identity

$$X^n = -a_{n-1}X^{n-1} - \cdots - a_0 \quad \text{when} \quad X = \alpha.$$

In this sense, $K[\alpha]$ behaves like a polynomial ring in α, subject to the relation given. (For more in this line, see section 8.3 and Exercise 10.22.) Suppose, for instance, that $m_{\alpha,K}(X) = X^3 + X + 1$. Then we have the relation $\alpha^3 = -\alpha - 1$, and we may reduce α^5 using the relation as follows:

$$\alpha^5 = \alpha^2 \cdot \alpha^3 = \alpha^2 \cdot (-\alpha - 1) = -\alpha^3 - \alpha^2 = -(-\alpha - 1) - \alpha^2 = -\alpha^2 + \alpha + 1.$$

8.2. The Ring Is a Field

Now we show that $K[\alpha]$ is not a proper subset of, but is in fact identical to, $K(\alpha)$. We do so by following the proof of the previous section, inserting an argument that each nonzero element of $K[\alpha]$ has a multiplicative inverse inside $K[\alpha]$.

---

[1] We use the nth root symbol $\sqrt[n]{a}$ to denote an arbitrary solution to $X^n - a = 0$, not necessarily real; if we need to specify a particular solution, we will provide additional information, such as a complex number approximation of the solution.

Theorem 8.9. *Let K be a subfield of $\mathbb{C}$ and α an algebraic number. Each element of $K(\alpha)$ has a unique representation as a sum of K-multiples of nonnegative powers of α, where the powers are less than $\deg_K(\alpha)$:*

$$K(\alpha) = \left\{ \sum_{i=0}^{\deg_K(\alpha)-1} k_i \alpha^i \ \bigg| \ k_0, k_1, \ldots, k_{(\deg_K(\alpha)-1)} \in K \right\}. \tag{8.4}$$

This representation will be called the reduced form in α *for elements of $K(\alpha)$. Hence $K[\alpha] = K(\alpha)$.*

Proof. We first observe that any arithmetic combination representing a number $\beta \in \mathbb{C}$ may be replaced by another arithmetic combination, also in α and representing the same β, in the form

$$\beta = \frac{\gamma}{\delta}, \qquad \gamma = \sum_{i=0}^{\deg_K(\alpha)-1} g_i \alpha^i, \qquad \delta = \sum_{i=0}^{\deg_K(\alpha)-1} d_i \alpha^i, \qquad g_i, d_i \in K.$$

(Exercise 10.21 suggests an algorithmic proof.)

We proceed by showing that δ has a multiplicative inverse $\delta^{-1} \in K[\alpha]$, using greatest common divisors. (See Exercise 10.23 for an alternate proof.) Since each element of $K[\alpha]$ may be expressed in reduced form, and since from the previous proof, we have that the product $\gamma \delta^{-1}$ of two numbers in reduced form may again be expressed in reduced form, the first claim of the theorem will have been shown. In particular, $K(\alpha) \subset K[\alpha]$, and the remainder of the theorem follows, since clearly $K[\alpha] \subset K(\alpha)$, giving us $K[\alpha] = K(\alpha)$.

We claim that for $0 \neq \delta \in K[\alpha]$, there exists $\delta^{-1} \in K[\alpha]$ such that $\delta\delta^{-1} = 1$. First let $f \in K[X]$ be a polynomial such that $f(\alpha) = \delta$, as follows: write δ as before in the reduced form

$$\delta = \sum_{i=0}^{\deg_K(\alpha)-1} d_i \alpha^i, \qquad d_i \in K,$$

and set

$$f = f(X) = \sum_{i=0}^{\deg_K(\alpha)-1} d_i X^i, \qquad d_i \in K.$$

The polynomial f clearly has degree less than $\deg_K(\alpha)$. Now observe that $\gcd(f, m_{\alpha,K}) = 1$, since $m_{\alpha,K}$ is irreducible and $\deg(f) < \deg_K(\alpha)$. By the Euclidean Algorithm

(Theorem 1.7), there exist polynomials $a, b \in K[X]$ such that

$$af + bm_{\alpha,K} = \gcd(f, m_{\alpha,K}) = 1. \tag{8.5}$$

Now $\delta^{-1} = a(\alpha) \in K[\alpha]$ is a multiplicative inverse for δ, as follows. Substituting α for X in equation (8.5), we have

$$1 = a(\alpha) f(\alpha) + b(\alpha) m_{\alpha,K}(\alpha) = \delta^{-1}\delta + b(\alpha) \cdot 0,$$

and we are done. $\qquad\qquad\qquad\qquad\qquad\qquad\qquad\qquad\qquad\qquad\qquad\square$

Example 8.10. We seek a multiplicative inverse for $\alpha^2 - 1$ in $\mathbb{Q}[\alpha]$, where $m_{\alpha,K} = X^3 - 2$.

We first write $\alpha^2 - 1$ and $m_{\alpha,K}$ as polynomials in X and execute the Euclidean Algorithm (Theorem 1.7) on $f = f(X) = X^2 - 1$ and $g = g(X) = X^3 - 2$, the steps of which are as follows:

$$
\begin{aligned}
(r_0 = q_1 r_1 + r_2): \quad & X^2 - 1 = (0)(X^3 - 2) + (X^2 - 1) \\
(r_1 = q_2 r_2 + r_3): \quad & X^3 - 2 = (X)(X^2 - 1) + (X - 2) \\
(r_2 = q_3 r_3 + r_4): \quad & X^2 - 1 = (X + 2)(X - 2) + (3) \\
(r_3 = q_4 r_4 + r_5): \quad & X - 2 = \left(\frac{1}{3}X - \frac{2}{3}\right)(3) + 0.
\end{aligned}
$$

Hence the last nonzero remainder is $r_4 = 3$. Dividing the next-to-last equation by 3 and working backwards (see Exercise 5.10), we may express 1 as a linear combination $af + bg$ of f and g with polynomials $a, b \in \mathbb{Q}[X]$:

$$1 = \frac{1}{3}(X^2 + 2X + 1) f - \frac{1}{3}(X + 2)g.$$

We read off the inverse from the polynomial $a = a(X) = \frac{1}{3}(X^2 + 2X + 1)$ by substituting α for X. We may check that $a(\alpha)$ is an inverse by multiplying $\alpha^2 - 1$ by the inverse and reducing, to yield 1:

$$\left(\frac{1}{3}\right)(\alpha^2 + 2\alpha + 1)(\alpha^2 - 1) = (\alpha^4 + 2\alpha^3 - 2\alpha - 1)(1/3) = 1.$$

Example 8.11. Now suppose that in the same field $\mathbb{Q}(\alpha)$ with $m_{\alpha,K} = X^3 - 2$, we wish to find the reduced form for $(\alpha + 1)/(\alpha^2 - 1)$. Using the inverse computed in the previous example, we multiply $(\alpha + 1) \cdot (1/3)(\alpha^2 + 2\alpha + 1)$ to find

$$\alpha^3/3 + \alpha^2 + \alpha + 1/3,$$

which, reduced modulo $m_{\alpha,\mathbb{Q}}$, is $\alpha^2 + \alpha + 1$. Hence the reduced form of $(\alpha + 1)/(\alpha^2 - 1)$ is $\alpha^2 + \alpha + 1$.

8.3. These Fields Are Isomorphic to Quotients of Polynomial Rings

Now we show how the field $K(\alpha)$ generated by α is isomorphic to a field defined only as a quotient field of the polynomial ring $K[X]$, where X is an indeterminate and the quotient is taken modulo the ideal generated by $m_{\alpha,K}(X)$. This quotient of a polynomial ring may be viewed as independent of – albeit isomorphic to a subfield of – $\mathbb{C}$, and it will be of use to us later in decomposing fields generated by several algebraic numbers.

Theorem 8.12 (Structure Theorem for Fields Generated by an Algebraic Number). *(Compare Theorems 16.5 and 35.1.) Let K be a subfield of $\mathbb{C}$ and α an algebraic number. The field $K(\alpha)$ generated over K by α is isomorphic to $K[X]/(m_{\alpha,K}(X))$.*

Proof. Define a ring homomorphism $\varphi \colon K[X] \to K[\alpha]$, called the *evaluation homomorphism*, by $\varphi(p) = p(\alpha)$ for each $p \in K[X]$ (see Exercise 10.17). The kernel of φ is precisely the set of polynomials associated to α, which is the ideal $(m_{\alpha,K}(X))$ (Proposition 7.2). By the First Isomorphism Theorem for Rings (Theorem 3.1), the ring homomorphism $\bar{\varphi} \colon K[X]/(m_{\alpha,K}(X)) \to K[\alpha]$ defined by

$$\bar{\varphi}(p + (m_{\alpha,K}(X))) = p(\alpha)$$

is one-to-one. Now each element of $K[\alpha]$ possesses a reduced form that is a sum of K-multiples of powers of α (Theorem 8.6). Viewing this sum as a polynomial $p(\alpha)$, we find that $\bar{\varphi}$ is onto, since $\bar{\varphi}(p + (m_{\alpha,K}(X))) = p(\alpha)$. Since $K[\alpha]$ is moreover a field (equal to $K(\alpha)$), every nonzero element possesses a multiplicative inverse, and the isomorphism property tells us that $K[X]/(m_{\alpha,K}(X))$ must therefore be a field also. Hence $\bar{\varphi}$ is an isomorphism of fields. $\square$

Remark 8.13. Much of what we have done so far can be generalized to a wider setting, say to general fields of characteristic 0 or even to arbitrary fields. As the text progresses, you will periodically be encouraged to determine which elements of the theory may be abstracted and then to prove the corresponding statements. You might begin by considering how this theorem may be generalized.

9. Reduced Forms in $\mathbb{Q}(\alpha)$: *Maple* and *Mathematica*

The package `AlgFields` contains routines that automate the calculation of inverses and reduced forms in generated fields. We will introduce this package in more detail in Chapter 3, but here we provide enough detail to work with reduced expressions.[2]

To work with a generated field, first declare the field with the command

    `FDeclareField(f1, {m})` | `FDeclareField[f1, {m}]`

where `m` is an irreducible polynomial of one variable, say α, and `f1` is the name we will give the generated field. `AlgFields` then creates the necessary data structures for $\mathbb{Q}(\alpha)$. Once `f1` is declared, the command

    `FInvert(expr, f1)` | `FInvert[expr, f1]`

finds the inverse of the polynomial `expr` in $\mathbb{Q}[\alpha]$, and

    `FSimplifyE(expr,f1)` | `FSimplifyE[expr,f1]`

reduces a polynomial `expr` in $\mathbb{Q}[\alpha]$.

Example 9.1. We seek a reduced form for $(\alpha^3 + \alpha)(\alpha^2) + 2\alpha - (\alpha + 1)$ (cf. Example 8.7), an inverse for $\alpha^2 - 1$ (cf. Example 8.10), and a reduced form for $(\alpha + 1)/(\alpha^2 - 1)$ (cf. Example 8.11) in $\mathbb{Q}(\alpha)$, where $m_{\alpha,\mathbb{Q}} = X^3 - 2$.

---

[2] *Maple*, in fact, already contains routines that handle the calculation of inverses and reduced forms in generated fields: these can be invoked with the commands `evala(Normal( ))` and `reduce`. Similarly, *Mathematica*, with its `NumberTheory`AlgebraicNumberFields`` package, provides commands for the simplification of algebraic expressions in generated fields. In place of these, the package `AlgFields` provides functions with a simpler interface that help collect information about the field extensions.

First we declare the field:

```
>  FDeclareField(K,[alpha^3-2]);

----Details of field K ----

  Algebraic Numbers: [alpha]

  Dimension over Q: 3

  Minimal Polynomials: [alpha^3-2]

  Root Approximations: [

      1.2599210498948731648]

----
```

```
In[1]:= FDeclareField[K, {α^3 − 2}]

----Details of field K ----

  Algebraic Numbers: {α}

  Dimension over Q: 3

  Minimal Polynomials: {-2 + α^3}

  Root Approximations: {

      1.2599210498948731648}

----
```

Then, to find a reduced form for $(\alpha^3 + \alpha)(\alpha^2) + 2\alpha - (\alpha + 1)$:

```
>  FSimplifyE((alpha^3+alpha)*
>    (alpha^2)+2*alpha-(alpha+1),K);

      α + 1 + 2 α²
```

```
In[2]:= FSimplifyE[(α^3 + α)(α^2) + 2α − (α + 1), K]

Out[2]= 1 + α + 2 α²
```

Now we obtain the inverse of $\alpha^2 - 1$:

```
>  FInvert(alpha^2-1,K);

      1     2       1
      ─ α² + ─ α + ─
      3     3       3
```

```
In[3]:= FInvert[α^2 − 1, K]

              1     2 α     α²
Out[3]=    ─ + ─── + ──
              3      3       3
```

Finally, a reduced form for $(\alpha + 1)/(\alpha^2 - 1)$:

```
>  FSimplifyE((alpha+1)*
>    FInvert(alpha^2-1,K),K);

      α² + α + 1
```

```
In[4]:= FSimplifyE[(α + 1) ∗ FInvert[α^2 − 1, K], K]

Out[4]= 1 + α + α²
```

10. Exercises

A.

10.1* Prove that $I_{\alpha,K}$ in Proposition 7.2 is an ideal.

10.2 Let $\alpha = \sqrt{2}$. Find $m_{\alpha,\mathbb{Q}}$ and $m_{\alpha,\mathbb{R}}$. Be sure to establish the irreducibility of the minimal polynomials.

10.3 Let $\alpha = 1 + \sqrt{2}$ and $\beta = \sqrt{\alpha}$. Check that $m_{\alpha,\mathbb{Q}}(X) = X^2 - 2X - 1$. Find $m_{\beta,\mathbb{Q}}$.

10.4 Suppose that $X^3 + X$ is associated to an algebraic number α, but X^2 is not. Find $m_{\alpha,\mathbb{Q}}$. Is α determined uniquely, that is, is there a single algebraic number α with these properties?

10.5 Suppose that $X^5 + 2X^3 + X^2 + X + 1$ is associated to an algebraic number α, but $X^4 + 2X^2 + 1$ is not. Find $m_{\alpha,\mathbb{Q}}$. Is α determined uniquely?

10.6 Cite an appropriate theorem in order to deduce that to any given monic, irreducible polynomial in $\mathbb{Q}[X]$ there exists an algebraic number to which it is associated.

10.7 Let $m_{\alpha,\mathbb{Q}}(X) = X^2 + X + 1$ and find reduced forms for α^3 and α^4 over $\mathbb{Q}$. Check your results with *Maple* or *Mathematica*.

10.8 Let $m_{\alpha,\mathbb{Q}}(X) = X^3 + X + 1$. Find reduced forms for $\alpha^2, \alpha^3, \alpha^5$, and α^8 over $\mathbb{Q}$.

10.9 Find the reduced form for $\alpha^3 + \alpha^2 + \alpha + 1$ over $\mathbb{Q}$, where α is

  (1) $\sqrt{2}$;
  (2) $-\sqrt{2}$;
  (3) $\sqrt[3]{20}$.
  (4) What is ambiguous if α is $\sqrt[3]{-1}$?
  (5) What is ambiguous if α is $\sqrt[3]{1}$?

10.10* Prove that $K[\alpha]$ is a commutative ring for K an arbitrary subfield of $\mathbb{C}$.

10.11* Prove that $K[\alpha]$ is a vector space over K for K an arbitrary subfield of $\mathbb{C}$.

10.12 We write $1/\sqrt{2}$ for the inverse of $\sqrt{2}$ in $\mathbb{Q}[\sqrt{2}]$. Express this inverse in reduced form as a polynomial in $\sqrt{2}$ with coefficients in $\mathbb{Q}$.

10.13 Let $\alpha = \sqrt[3]{7}$. Find the inverse of each of α, α^2, and $\alpha + \alpha^2$, expressed in reduced form as polynomials in α with coefficients in $\mathbb{Q}$.

10.14 Using inverses, rewrite $\alpha^2/(1+\alpha^3)$ in reduced form over $\mathbb{Q}$, where $m_{\alpha,\mathbb{Q}} = X^2 + X + 1$.

10.15 Simplify into reduced form the expression $(1 + 2\alpha + 3\alpha^2)/(1+\alpha)^{15}$, where $m_{\alpha,\mathbb{Q}} = X^3 + X + 1$.

10.16 Find quotients of polynomial rings that are isomorphic to $\mathbb{Q}(\alpha)$ for each of the following:

   (1) $\alpha = \sqrt{2}$;

   (2) $\alpha = -\sqrt{2}$;

   (3) $\alpha = \sqrt[3]{3}$;

   (4) α is a root of $X^2 + X + 1$.

10.17* Prove that the evaluation homomorphism defined in Theorem 8.12 is a homomorphism.

10.18 Prove that for any fields $F \subset K$, $[K : F] = 1$ if and only if $K = F$.

B.

10.19 Let K be a subfield of $\mathbb{C}$. Prove that $K[\alpha]$ is isomorphic to $K[\beta]$ *as vector spaces over K* if and only if $\deg_K(\alpha) = \deg_K(\beta)$.

10.20* Prove that each number $\beta \in \mathbb{C}$ represented by an arithmetic combination in α over K without quotients is also represented by an arithmetic combination in α over K in polynomial form: $\beta = \sum_{i=0}^{n} c_i \alpha^i$, $n \geq 0$, $c_i \in K$. (Hint: For an algorithmic proof, define the *number of operations* of an arithmetic combination to be the number of field operations in the combination. This number may be zero; consider the arithmetic combinations α and 2. Then proceed by induction.)

10.21* Prove that each number $\beta \in \mathbb{C}$ represented by an arithmetic combination in α over K is also represented by an arithmetic combination in α over K in the form of a quotient of polynomials in α, each of degree less than $\deg_K(\alpha)$. (Hint: Proceed by induction as in the previous exercise, handling the degree condition at the end using Theorem 8.6.)

10.22* Let F be a field and X an indeterminate. Define a *formal algebraic element over F*, denoted by α, to be the coset $X + (p)$ in the quotient field $F_\alpha = F[X]/(p)$, where $p \in F[X]$ is some monic, irreducible polynomial $p \in F[X]$. Suppose now that $F = \mathbb{Q}$. Prove that F_α is isomorphic to at least one subfield of $\mathbb{C}$ and that under any such isomorphism, $X + (p)$

is sent to a root of p in $\mathbb{C}$. Is the study of algebraic numbers the same as the study of formal algebraic elements?

10.23* Let K be a subfield of $\mathbb{C}$. Prove the existence of inverses in $K[\alpha] = K(\alpha)$ using the following alternative outline. Let $\beta \in K[\alpha]$ with $\beta \neq 0$; we seek a multiplicative inverse in $K[\alpha]$ for β. Consider the function $f: K[\alpha] \to K[\alpha]$ given by $f(\gamma) = \beta \cdot \gamma$. (a) Show that f is a linear transformation on the vector space $K[\alpha]$; (b) use the fact that $K[\alpha]$ has finite dimension to prove that f is one-to-one if and only if f is onto; (c) show that f is one-to-one because β and γ lie in $K(\alpha) \subset \mathbb{C}$, which is an integral domain, having no zero-divisors; (d) conclude that $f^{-1}(1)$ exists and is β^{-1}.

10.24 Make Exercise 10.23 computationally explicit by showing that a system of linear equations may be constructed, the solution to which provides the coefficients of the inverse in the basis $1, \alpha, \alpha^2, \ldots, \alpha^{\deg_K(\alpha)-1}$, as follows: (a) express β as linear combination of the basis elements, as in $\beta = \sum_{i=0}^{\deg_K(\alpha)-1} b_i \alpha^i$; (b) let an arbitrary element of $K[\alpha]$ be written in the same basis, with indeterminate coefficients g_i: $\gamma = \sum_{i=0}^{\deg_K(\alpha)-1} g_i \alpha^i$; (c) show that finding the inverse of β is equivalent to finding the g_i such that

$$\left(\sum_{i=0}^{\deg_K(\alpha)-1} g_i \alpha^i \right) \left(\sum_{i=0}^{\deg_K(\alpha)-1} b_i \alpha^i \right) = 1;$$

(d) show further that the product on the left-hand side can be expressed in reduced form modulo $m_{\alpha,K}$; (e) show then that in the reduced form, the coefficients of the α^i are linear expressions in the variables g_i with coefficients coming from K; (f) note that for the two sides of the equation to be equal, coefficients of like powers of α must be equal, so that we have a system of linear equations that may be solved to find the inverse of β; (g) show that Gaussian elimination, or some other algorithm of linear algebra, finishes a constructive method of solution.

10.25 Use Exercise 10.24 to determine the linear equations necessary to solve for the inverse of $\alpha^2 + \alpha + 2$ when $m_{\alpha,\mathbb{Q}} = X^4 + X^2 + 5$.

10.26 Show that not every quotient ring of $\mathbb{Q}[X]$ is isomorphic to $\mathbb{Q}(\alpha)$ for some algebraic number α. (Hint: Consider a quotient ring by a principal ideal (fg) for two nonconstant polynomials f and g in $\mathbb{Q}[X]$.)

10.27 Prove that $K_1 = \mathbb{Q}(\sqrt{2})$ and $K_2 = \mathbb{Q}(-\sqrt{2})$ are equal, hence isomorphic under the identity map. Now let $g_1 : \mathbb{Q}[X]/(X^2 - 2) \to K_1$ and $g_2 : \mathbb{Q}[X]/(X^2 - 2) \to K_2$ be the corresponding isomorphisms in the proof of Theorem 8.12. Is $g_2 \circ g_1^{-1}$ equal to the identity map?

10.28 Let α be an algebraic number and K a field containing $\mathbb{Q}$. Prove that $(m_{\alpha,K}(X))$ is a maximal ideal in $K[X]$ and hence that $K[X]/(m_{\alpha,K}(X))$ is a field. Use the evaluation homomorphism to prove, without Theorem 8.9, that $K[\alpha]$ is a field and hence that $K[\alpha] = K(\alpha)$.

CHAPTER THREE

Working with Algebraic Numbers, Field Extensions, and Minimal Polynomials

Over any field $K \subset \mathbb{C}$, we have associated to each algebraic number α a polynomial $m_{\alpha,K}$ and a field $K(\alpha)$. In this chapter, we consider to what extent these objects are unique, that is, whether a given polynomial or field might be generated by other algebraic numbers β as well. In the first section, we determine all of the algebraic numbers to which a minimal polynomial is associated. In the second, we find that if a field is generated by an algebraic number, it is generated by each of an infinite number of algebraic numbers, and we introduce a notion of size of a generated field. Finally, in section 14 we introduce tools for exploring fields generated by an algebraic number.

11. Minimal Polynomials Are Associated to Which Algebraic Numbers?

We seek to describe the relationship between algebraic numbers and minimal polynomials as precisely as possible, and we begin by observing how many roots a polynomial may have in a field. We will then relate these roots to linear factors of the polynomial over the complex numbers, and we finally consider what we may say about minimal – that is, irreducible – polynomials.

11.1. A Polynomial of Degree n Has at Most n Roots in Any Field Extension

We first present a theorem over arbitrary fields that bounds the number of roots a polynomial may have in a field.

Theorem 11.1. *Let K be a field, $p \in K[X]$ a nonzero polynomial, $n = \deg(p)$, and L a field containing K. Then p has no more than n roots in L.*

Proof. We prove the statement by induction on $\deg(p)$. If $\deg(p) = 0$, p is a nonzero constant polynomial and hence has no roots. Now suppose that the statement holds for degree n, and let p be a polynomial of degree $n + 1$. If p has no roots in L, then we are done; otherwise let $\alpha \in L$ be a root of p.

We claim that $p = (X - \alpha)q$ for some polynomial $q \in L[X]$, as follows. By the Division Algorithm (Theorem 1.5), we may divide $X - \alpha$ into p, yielding

$$p = (X - \alpha)q + r \tag{11.1}$$

for a polynomial $r \in L[X]$ which is either zero or of degree less than $\deg(X - \alpha) = 1$. Hence r is a constant polynomial. Evaluating both sides of equation (11.1) at $X = \alpha$, we obtain $0 = 0 + r(\alpha)$. Therefore $r = 0$, and the claim is proved.

Now suppose that $\beta \in L$ is a root of p. Then $0 = p(\beta) = (\beta - \alpha)q(\beta)$. This is an equation of elements of L, and since no product of two elements of L is zero unless at least one of them is zero, we deduce that either $\beta = \alpha$ or β is a root of q. Because the polynomial q has degree n, we may use the inductive hypothesis to conclude that the number of roots of p in L is at most $n + 1$. $\qquad\square$

Hence the number of roots in L of a nonzero polynomial $p \in K[X]$ with $K \subset L$ is bounded below by zero and above by $\deg(p)$. In this generality, the bounds are the best possible. The upper bound is reached, for instance, when $L = \mathbb{C}$ and the polynomial contains no multiple roots. The lower bound is reached for $K = L = \mathbb{Q}$ and any irreducible polynomial of degree greater than 1. Intermediate situations occur as L moves between the two extremes $\mathbb{Q}$ and $\mathbb{C}$.

11.2. A Polynomial of Degree n Factors into n Linear Factors over $\mathbb{C}$

We want to show that for any minimal polynomial p of degree n, there exist precisely n algebraic numbers for which p is the minimal polynomial. Our method will be to establish a correspondence between roots $\alpha \in \mathbb{C}$ of p and linear factors $X - \alpha$ of p. We first prove that any polynomial $p \in K[X]$, $K \subset \mathbb{C}$, factors into at most $n = \deg(p)$ linear factors over $\mathbb{C}$, some of which may be identical. Then we establish that since minimal polynomials are irreducible, all n of these factors are distinct and, if $p \in \mathbb{Q}[X]$, the corresponding roots α are algebraic numbers.

Theorem 11.2. *Let K be a subfield of $\mathbb{C}$. Let $p \in K[X]$ be a nonzero polynomial and $n = \deg(p)$. Then there exist complex numbers $\alpha_1, \alpha_2, \dots, \alpha_k \in \mathbb{C}$, and positive integers*

$m_1, m_2, \ldots, m_k, k \le n$, *such that*

$$p = p(X) = c_n(X - \alpha_1)^{m_1}(X - \alpha_2)^{m_2} \cdots (X - \alpha_k)^{m_k},$$

where $c_n \in K$ is the leading coefficient of p. The set of ordered pairs $\{(\alpha_i, m_i)\}$ is uniquely determined by p. We say that root α_i has multiplicity m_i.

While we could use Theorem 1.8 and Corollary 1.14, we present a more algorithmic proof.

Proof. We use the Fundamental Theorem of Algebra (Theorem 1.13) to prove the theorem by induction on $\deg(p)$. If p has degree zero, then p is a nonzero constant and hence has the unique representation given in the theorem with $k = 0$. Now suppose that for every subfield K of $\mathbb{C}$, each nonzero polynomial $q \in K[X]$ of degree at most n has the unique representation given in the theorem, and let $p \in K[X]$ be a polynomial with $\deg(p) = n + 1$.

By the Fundamental Theorem of Algebra, p must have at least one root in $\mathbb{C}$, say α. By the Division Algorithm (Theorem 1.5), we may divide p by $X - \alpha$ over the field $\mathbb{C}$. As in the proof of Theorem 11.1, $X - \alpha$ must divide p evenly, so that $p = (X - \alpha)q$ for a polynomial q of degree n in $\mathbb{C}[X]$. By induction, q has the representation given in the theorem; p therefore has a similar representation with a set of pairs (α_i, m_i) of roots and multiplicities that is either (a) one pair larger, by the introduction of a new root with pair $(\alpha_{k+1}, 1)$, or (b) the same size but with one pair altered, namely by the addition of 1 to the multiplicity of a root of q.

The representation for p is unique, as follows. Suppose there exist two such sets $\{(\alpha_i, m_i)\}$, $\{(\alpha_i', m_i')\}$. First, the sets of zeros $\{\alpha_i\}$, $\{\alpha_i'\}$ must be the same for each set, since each α_i is a root by definition, and if α is a root and p has the decomposition as in the theorem, then by the zero product property, $\alpha = \alpha_i$ for some i. Hence we may assume without loss of generality that $\alpha_i = \alpha_i'$ for each i.

Now consider the polynomial $\prod_i (X - \alpha_i)^{\min(m_i, m_i')}$. If it is equal to p, then by counting degrees, we see that $m_i = m_i'$ for each i and the two representations for p are identical. Otherwise, divide each representation of p by $\prod_i (X - \alpha_i)^{\min(m_i, m_i')}$. Then we have two distinct representations of a polynomial of smaller degree that differ in factorization, contradicting the inductive hypothesis. Hence both representations are the same. $\qquad\square$

11.3. Minimal Polynomials Are Minimal Polynomials for n Distinct Algebraic Numbers

Theorem 11.3. *Let K be a subfield of $\mathbb{C}$ and let $p \in K[X]$ be a polynomial that is irreducible over K. Then the multiplicity (as in Theorem 11.2) of each root is 1.*

Proof. Let $\{\alpha_i\}$ be the set of roots of p as in Theorem 11.2, and suppose that the multiplicity of some root α_i is at least two. Consider the derivative p' of p. It is an exercise (13.11) to show that if $p \in K[X]$ then $p' \in K[X]$. Since $\alpha_i \in \mathbb{C}$ has multiplicity at least two in p, then $p = (X - \alpha_i)^2 q$ for some polynomial q with coefficients in $\mathbb{C}$. Then

$$p'(X) = 2(X - \alpha_i)q(X) + (X - \alpha_i)^2 q'(X) = (X - \alpha_i)h(X)$$

for some nonzero polynomial $h(X) = 2q(X) + (X - \alpha_i)q'(X) \in \mathbb{C}[X]$. Hence $p'(\alpha) = 0$.

Now the greatest common divisor $\gcd(p, p')$ of p and p' may be expressed as a $K[X]$-linear combination $zp + wp'$, $z, w \in K[X]$, of p and p' by the Euclidean Algorithm (Theorem 1.7). But because p is irreducible and p' is of degree strictly smaller than the degree of p, $\gcd(p, p') = 1$. Hence $1 = zp + wp'$, and substituting α for the indeterminate X, we derive $1 = 0$, a contradiction. $\square$

Corollary 11.4. *If $p \in \mathbb{Q}[X]$ is an irreducible polynomial over $\mathbb{Q}$, then p has $n = \deg(p)$ distinct roots, all of which are algebraic numbers.*

Definition 11.5. Let K be a subfield of $\mathbb{C}$ and α and β algebraic numbers. We say that α and β are *conjugate over K* if α and β share the same minimal polynomial over K, that is, if $m_{\alpha,K} = m_{\beta,K}$. If $K = \mathbb{Q}$, then we simply say that α and β are *conjugate*.

Hence for every subfield K of $\mathbb{C}$, each algebraic number α has $(\deg_K(\alpha) - 1)$ distinct other conjugates over K.

12. Which Algebraic Numbers Generate a Generated Field?

We now take up the question of which algebraic numbers generate a given generated field, that is, given $K(\alpha)$, which $\beta \in K(\alpha)$ give $K(\alpha) = K(\beta)$? By definition, the only algebraic numbers that may generate a field $K(\alpha)$ over K are those that are inside it, that is, those that are represented by arithmetic combinations of α with coefficients in K. Now if $\beta \in K(\alpha)$, then any arithmetic combination of β is, by substitution, an arithmetic combination of α. Hence $K(\beta) \subset K(\alpha)$ – but perhaps $K(\beta)$ is a proper subfield of $K(\alpha)$. How may we tell?

12.1. Degrees of Minimal Polynomials of Algebraic Numbers Generating a Given Field

We approach our problem by associating a size to $K(\alpha)$ and $K(\beta)$, that of their vector space dimensions over K, in order to have a test for equality.

Theorem 12.1. *Let α be an algebraic number, K a subfield of $\mathbb{C}$, and β an algebraic[1] number contained in the field $K(\alpha)$. Then the following are equivalent:*

- *β generates the field $K(\alpha)$ over K, that is, $K(\beta) = K(\alpha)$;*
- *$[K(\beta) : K] = [K(\alpha) : K]$;*
- *$\deg_K(\alpha) = \deg_K(\beta)$.*

Proof. By Theorems 8.6 and 8.9, $K(\alpha)$ and $K(\beta)$ are vector spaces over K of respective dimensions $\deg_K(\alpha)$ and $\deg_K(\beta)$. From the theory of linear algebra, a subspace of a finite-dimensional vector space of the same dimension as the whole space must in fact be the whole space. Hence

$$[K(\beta) : K] = [K(\alpha) : K] \quad \text{if and only if} \quad K(\beta) = K(\alpha). \qquad \square$$

Example 12.2. Let $\alpha = \sqrt{2}$, $\beta = \sqrt{2} + 1 = \alpha + 1$, and $K = \mathbb{Q}$. It is not difficult to see that any arithmetic combination of α can be rewritten as an arithmetic combination of β, simply by replacing α by $\beta - 1$; similarly, one can rewrite any arithmetic combination of β as an arithmetic combination of α. Therefore we have that $\mathbb{Q}(\alpha) = \mathbb{Q}(\beta)$. To check our result using the theorem, note that $m_\alpha = X^2 - 2$ while $m_\beta = X^2 - 2X - 1$, and both have the same degree.

Example 12.3. Let α have minimal polynomial $m_{\alpha,\mathbb{Q}} = X^4 - 10X^2 + 20$. The arithmetic combination $\beta = 20 - 4\alpha^2 \in \mathbb{Q}(\alpha)$ does not generate $\mathbb{Q}(\alpha)$, because there exists a polynomial in $\mathbb{Q}[X]$ of degree less than 4 that is associated to β, namely, $X^2 - 80$, which we verify as follows:

$$(20 - 4\alpha^2)^2 - 80 = 16\alpha^4 - 160\alpha^2 + 400 - 80 = 16(\alpha^4 - 10\alpha^2 + 20) + 80 - 80 = 0.$$

---

[1] See Theorem 12.12 for a hypothesis on K that implies this condition.

Since $X^2 - 80$ is irreducible over $\mathbb{Q}$, we have that $m_\beta = X^2 - 80$ and so $[\mathbb{Q}(\beta) : \mathbb{Q}] = 2 < 4 = [\mathbb{Q}(\alpha) : \mathbb{Q}]$.

We have not yet established which elements of $\mathbb{Q}(\alpha)$ are algebraic numbers, and we have not established a method for finding $m_{\beta,K}$, given $m_{\alpha,K}$, for $\beta \in K(\alpha)$. We address the first question in section 12.4 and the second in section 14.

12.2. If an Algebraic Number Generates a Field, So Do Its Affine Translations

The preceding theorem gives a criterion for algebraic numbers other than α to generate $K(\alpha)$ over K, but the result does not indicate how to find such a β. We show, however, that affine translations of α – that is, elements $a\alpha + b$ for $a, b \in \mathbb{Q}$ – give other generators of the field, and that these alternate generators are themselves algebraic numbers.

Theorem 12.4. *Let K be a subfield of $\mathbb{C}$, and let α be an algebraic number. Set $\beta = a\alpha + b$ for $a, b \in \mathbb{Q}$ with $a \neq 0$. Then β is an algebraic number that generates $K(\alpha)$ over K, that is, $K(\beta) = K(\alpha)$.*

Proof. Suppose $p \in \mathbb{Q}[X]$ is a polynomial associated to α. Then the polynomial $q \in \mathbb{Q}[X]$ defined by $q(X) = p((X - b)/a)$ is associated to β. Similarly, if $q \in \mathbb{Q}[X]$ is a polynomial associated to β, then the polynomial $p \in \mathbb{Q}[X]$ defined by $p(X) = q(aX + b)$ is associated to α.

Now let $p = m_{\alpha,\mathbb{Q}}$. Since minimal polynomials are irreducible (Proposition 7.2), p is irreducible. If $q = q_1 q_2$ for polynomials $q_1, q_2 \in \mathbb{Q}[X]$, then $p = p_1 p_2$ where $p_i(X) = q_i(aX + b)$ for $i = 1, 2$. Hence q is irreducible also. Now p and q are irreducible polynomials of the same degree, and so the minimal polynomials of α and β over $\mathbb{Q}$ have the same degree. Therefore the generated fields have the same dimensions over $\mathbb{Q}$: $[\mathbb{Q}(\alpha) : \mathbb{Q}] = [\mathbb{Q}(\beta) : \mathbb{Q}]$ (Theorem 8.6), and since $\beta \in \mathbb{Q}(\alpha)$, we may conclude that β generates the same field as α: $\mathbb{Q}(\alpha) = \mathbb{Q}(\beta)$ (Theorem 12.1). Therefore α may be represented by an arithmetic combination of β, and β may be represented by an arithmetic combination of α, each with coefficients in $\mathbb{Q}$. But then $K(\alpha) \subset K(\beta)$ and $K(\beta) \subset K(\alpha)$, from which we conclude that $K(\alpha) = K(\beta)$. $\square$

Example 12.5. Let $\alpha = \sqrt[3]{2}$, $\beta = 2\alpha + 3$, and $K = \mathbb{Q}$. Then Theorem 12.4 tells us that β is an algebraic number. We use the method of the proof to find $m_{\beta,\mathbb{Q}}$. By evaluating $m_{\alpha,\mathbb{Q}}$ at

$(X - 3)/2$, we find a multiple of $m_{\beta,\mathbb{Q}}$:

$$q = m_{\alpha,\mathbb{Q}}((X - 3)/2) = ((X - 3)/2)^3 - 2 = (1/8)(X - 3)^3 - 2$$

$$= (1/8)(X^3 - 9X^2 + 27X - 27) - 2$$

$$= (1/8)X^3 - (9/8)X^2 + (27/8)X + (-43/8).$$

Hence $m_{\beta,\mathbb{Q}} = 8q$, and $\mathbb{Q}(\alpha) = \mathbb{Q}(\beta)$.

Example 12.6. Using *Maple* and *Mathematica*, we find the polynomial q of the proof of Theorem 12.4 by substitution. We keep the notation of the previous example.

```
>  q:=subs(x=(x-3)/2,x^3-2);

        q := (1/2 x - 3/2)³ - 2
```

```
In[1]:= q = (X^3 - 2) /. {X → (X - 3)/2}

Out[1]= -2 + 1/8 (-3 + X)³
```

Following this expression by commands for expansion and factorization (expand and factor in *Maple*, Expand and Factor in *Mathematica*), we reach

```
>  q2:=expand(q);

    q2 := 1/8 x³ - 9/8 x² + 27/8 x - 43/8
```

```
In[2]:= q2 = Expand[q]

Out[2]= -43/8 + 27 X/8 - 9 X²/8 + X³/8
```

```
>  factor(q2);

    1/8 x³ - 9/8 x² + 27/8 x - 43/8
```

```
In[3]:= Factor[q2]

Out[3]= 1/8 (-43 + 27 X - 9 X² + X³)
```

We have confirmed that q is irreducible.

12.3. Degrees of Minimal Polynomials Divide the Dimension of an Enclosing Field

Degrees of minimal polynomials over K of algebraic numbers $\beta \in K(\alpha)$ are multiplicatively related to the degree $[K(\alpha) : K]$. We first illustrate this phenomenon with two examples, where, for the first time, we use a factorization of polynomials not in $\mathbb{Q}[X]$, but in $\mathbb{Q}(\beta)[X]$. Factoring polynomials over extension fields of $\mathbb{Q}$ is certainly nontrivial (see, for instance, [40]), and later, in section 14, we will introduce computational tools to perform such factorizations.

Example 12.7. Let $K = \mathbb{Q}$, and consider $\mathbb{Q}(\alpha)$, where $m_{\alpha,\mathbb{Q}} = X^4 - 10X^2 + 20$. In Example 12.3, we found that $\mathbb{Q}(\beta) \subsetneq \mathbb{Q}(\alpha)$, where $\beta = 20 - 4\alpha^2$; we determined that $\mathbb{Q}(\beta)$ is a proper subfield of $\mathbb{Q}(\alpha)$ with $[\mathbb{Q}(\beta) : \mathbb{Q}] = \deg(m_{\beta,\mathbb{Q}}) = 2$. Now since $m_{\alpha,\mathbb{Q}}$ has α as a root, some factor in the factorization of $m_{\alpha,\mathbb{Q}}$ over $\mathbb{Q}(\beta)$ must be the minimal polynomial of α over $\mathbb{Q}(\beta)$, denoted $m_{\alpha,\mathbb{Q}(\beta)}$. Examining the factorization

$$X^4 - 10X^2 + 20 = (X^2 - \frac{1}{4}\beta - 5) \cdot (X^2 + \frac{1}{4}\beta - 5) \in \mathbb{Q}(\beta)[X]$$

$m_{\alpha,\mathbb{Q}}$ over $\mathbb{Q}(\beta)$, we see that the degree of $m_{\alpha,\mathbb{Q}(\beta)}$ must be 2. From this example, we find that $\deg(m_{\beta,\mathbb{Q}})$ divides $[\mathbb{Q}(\alpha) : \mathbb{Q}] = 4$ and $\deg(m_{\alpha,\mathbb{Q}(\beta)})$ divides 4 as well.

Example 12.8. Let $K = \mathbb{Q}$ and consider $\mathbb{Q}(\alpha)$, where $\alpha = \sqrt[6]{2}$ is one of the two real sixth roots of 2. Let $\beta = \alpha^2 = \sqrt[3]{2}$ be the real third root. Then we have $[\mathbb{Q}(\alpha) : \mathbb{Q}] = 6$ since the minimal polynomial $m_{\alpha,\mathbb{Q}} = X^6 - 2$ is irreducible, and the minimal polynomial $m_{\beta,\mathbb{Q}} = X^3 - 2$ is similarly irreducible, so $[\mathbb{Q}(\beta) : \mathbb{Q}] = 3$. That this degree divides $[\mathbb{Q}(\alpha) : \mathbb{Q}] = 6$ is similarly no coincidence.

Theorem 12.9. *Let K be a subfield of $\mathbb{C}$. Let α be an algebraic number and $\beta \in K(\alpha)$ another algebraic number. Then, as integers, $\deg_K(\beta) = [K(\beta) : K]$ divides $\deg_K(\alpha) = [K(\alpha) : K]$. In fact,*

$$\deg_K(\alpha) = [K(\alpha) : K] = [K(\alpha) : K(\beta)][K(\beta) : K] = \deg_{K(\beta)}(\alpha) \deg_K(\beta).$$

As a visual aid to understanding a tower of field extensions, a subfield diagram is commonly drawn, with the degrees indicated next to the connecting segments. For Theorem 12.9, the diagram is

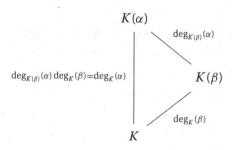

Proof. First observe that $K(\alpha)$ is precisely the field generated by α over the field $K(\beta)$, as follows. Since $\beta \in K(\alpha) = K[\alpha]$, $K(\beta) \subset K[\alpha]$. Furthermore, $K[\alpha]$ is a vector space over the field $K(\beta)$ (Theorem 8.6). Then, this vector space $K[\alpha]$ is in fact a field, $K(\alpha)$, with dimension over $K(\beta)$ equal to the degree of the minimal polynomial $m_{\alpha, K(\beta)}$ (Theorem 8.9).

In what follows, we will see the dimension of $K(\alpha)$ over K in two pieces: the dimension of $K(\alpha)$ over the subfield $K(\beta)$ and the dimension of $K(\beta)$ over K.

We claim first that $[K(\alpha) : K(\beta)]$ must be finite. The polynomial

$$m_{\alpha, K}(X) \in K[X] \subset K(\beta)[X]$$

is associated to α and has finite degree, say n. The minimal polynomial $m_{\alpha, K(\beta)}$ for α over $K(\beta)$ divides $m_{\alpha, K}$ (why?), so $\deg_{K(\beta)}(\alpha) \leq \deg_K(\alpha)$. Let $\{\gamma_1, \gamma_2, \ldots, \gamma_m\}$ be a basis for $K(\alpha)$ over $K(\beta)$, and let $\{v_1, v_2, \ldots, v_n\}$ be a basis for $K(\beta)$ over K. We will show that the set of products $\{\gamma_i v_j\}$ as i runs from 1 to m and j runs from 1 to n is a basis for $K(\alpha)$ over K and hence that $[K(\alpha) : K] = [K(\alpha) : K(\beta)][K(\beta) : K]$, which implies the result.

Let δ be an arbitrary element of $K(\alpha)$. Since $\{\gamma_1, \gamma_2, \ldots, \gamma_m\}$ is a basis for $K(\alpha)$ over $K(\beta)$, δ may be written as a linear combination of $\gamma_1, \gamma_2, \ldots, \gamma_m$ with coefficients in $K(\beta)$, say

$$\delta = \sum_{i=1}^{m} c_i \gamma_i, \quad c_i \in K(\beta).$$

Then, since $\{v_1, v_2, \ldots, v_n\}$ is a basis for $K(\beta)$ over K, each of these coefficients c_i may be written as a linear combination of $v_1, v_2, \ldots, v_n$ with coefficients in K, say as

$$c_i = \sum_{j=1}^{n} c_{i,j} v_j, \quad c_{i,j} \in K, \quad i = 1, \ldots, m.$$

Multiplying out and collecting terms, we see that δ may then be written as a linear combination of the elements of $\{\gamma_i v_j\}$ with coefficients in K:

$$\delta = \sum_{i=1}^{m} c_i \gamma_i = \sum_{i=1}^{m} \left(\sum_{j=1}^{n} c_{i,j} v_j \right) \gamma_i = \sum_{i=1}^{m} \sum_{j=1}^{n} c_{i,j} (\gamma_i v_j).$$

Since δ was arbitrarily chosen in $K(\alpha)$, we have established that the set $\{\gamma_i v_j\}$ spans $K(\alpha)$ over K.

Now we show that this set $\{\gamma_i v_j\}$ is linearly independent, by contradiction. Suppose that the set $\{\gamma_i v_j\}$ is linearly dependent. Then, viewing a linear dependence relation

$$\sum_{i,j} c_{i,j} \gamma_i v_j = 0, \quad \text{with not all } c_{i,j} = 0,$$

as a linear dependence of the γ_i with coefficients in $K(\beta)$ by rewriting it as

$$\sum_i \left(\sum_j c_{i,j} v_j \right) \gamma_i = 0,$$

we see that $\{\gamma_1, \gamma_2, \ldots, \gamma_m\}$ is not a basis for $K(\alpha)$ over $K(\beta)$, which is a contradiction.

We have therefore established that the set $\{\gamma_i v_j\}$ is a basis for $K(\alpha)$ over K, and hence that $[K(\alpha) : K] = mn = [K(\alpha) : K(\beta)][K(\beta) : K]$, proving the theorem. $\square$

Example 12.10. Let $\alpha = \sqrt[4]{2}$. Since $[\mathbb{Q}(\alpha) : \mathbb{Q}] = 4$, the only possibilities for the dimension of a subfield $\mathbb{Q}(\beta) \subset \mathbb{Q}(\alpha)$ over $\mathbb{Q}$ are 1, 2, and 4. Given $\beta \in \mathbb{Q}(\alpha)$, we may determine this dimension by calculating the degree of $m_{\beta,\mathbb{Q}}$. (See the computation section 14 at the end of this chapter for tools to do so.) By appropriately choosing β, we show that each of the possibilities 1, 2, 4 of $[\mathbb{Q}(\beta) : \mathbb{Q}]$ occurs.

If $\beta = 1$, then $m_{\beta,\mathbb{Q}} = X - 1$, of degree 1, and $\mathbb{Q}(\beta) = \mathbb{Q}$.

If $\beta = \sqrt{2} - 1 = \alpha^2 - 1$, then $m_{\beta,\mathbb{Q}} = X^2 + 2X - 1$, of degree 2, and $\mathbb{Q} \subsetneq \mathbb{Q}(\beta) \subsetneq \mathbb{Q}(\alpha)$.

Finally, if $\beta = \alpha + 1$, then by Theorem 12.4 and its proof, $m_{\beta,\mathbb{Q}} = X^4 - 4X^3 + 6X^2 - 4X - 1$, of degree 4, and $\mathbb{Q}(\beta) = \mathbb{Q}(\alpha)$.

Remark 12.11. The preceding theorem may be generalized to a statement about arbitrary field extensions rather than fields generated over another by an algebraic number, and in doing so, the portion of the result applying only to degrees of the field extensions is then placed in a simpler context. This generalization is Exercise 13.8, and this exercise will be frequently used in what follows.

12.4. The Set of Algebraic Numbers Is Closed Under Field Operations

We close this section by showing that adjoining an algebraic number to a field of algebraic numbers expands the field only by algebraic numbers. Equivalently, the sum, difference, product, or quotient of algebraic numbers is again an algebraic number. For example, if α and β are algebraic numbers, then we may show that $\alpha + \beta$ is an algebraic number as follows. First, since $\mathbb{Q}$ is certainly a field of algebraic numbers, $K = \mathbb{Q}(\alpha)$ is a field of algebraic numbers. Then $K(\beta)$ is a field of algebraic numbers as well. Since $\alpha + \beta \in K(\beta)$, we deduce that $\alpha + \beta$ is an algebraic number.

Theorem 12.12. *Suppose that K is a field of algebraic numbers, and let α be an algebraic number. Then each $\beta \in K(\alpha)$ is an algebraic number.*

Proof. First assume that $[K : \mathbb{Q}]$ is finite. Since $[K(\alpha) : K] = \deg m_{\alpha,K}(X)$, then by the Tower Law (Exercise 13.8), $[K(\alpha) : \mathbb{Q}]$ is finite. Since $\mathbb{Q}(\beta) \subset K(\alpha)$, we must have that $[\mathbb{Q}(\beta) : \mathbb{Q}]$ is finite as well. Hence the set $\{\beta^i\}$ for i ranging through the nonnegative integers must be linearly dependent over $\mathbb{Q}$, and a dependence relation gives us a polynomial over $\mathbb{Q}$ associated to β.

Now we consider the case in which $[K : \mathbb{Q}]$ is infinite. Let $\beta \in K(\alpha)$. Then β is represented by an arithmetic combination of α over K. In particular, this combination involves only a finite number of elements $\{k_1, k_2, \ldots, k_n\}$ of K, each algebraic. Let $K_0 = \mathbb{Q}$ and recursively define fields $K_{i+1} = K_i(k_{i+1})$ for $0 \leq i < n$. Applying the previous paragraph and the Tower Law (Exercise 13.8), we have that each $K_{i+1}, 0 \leq i < n$, is a field of algebraic numbers with $[K_{i+1} : \mathbb{Q}]$ finite. In particular, $[K_n : \mathbb{Q}]$ is finite. Applying the Tower Law again, we have that $K_n(\alpha)$ is a field of algebraic numbers with $[K_n(\alpha) : \mathbb{Q}]$ finite. Because of its representation as the arithmetic combination, $\beta \in K_n(\alpha)$, and we are in the situation of the previous paragraph, whereby we are done. $\square$

This result enables us to consider finite towers of fields, each one obtained from the previous by adjoining an algebraic number, so that each field in the tower consists entirely of algebraic numbers. In Chapter 4, we take up the general question of fields generated by more than one algebraic number.

Note that the proof of Theorem 12.12 does not tell us how to find associated polynomials to $\beta \in K(\alpha)$ for K a generated field, or further how to find minimal polynomials for such β. We take up this question, together with the question of finding generating algebraic numbers for generated fields, in section 14.

13. Exercise Set 1

A.

13.1* Prove that $p \in K[X]$ has a root in K if and only if the factorization of p over K contains at least one linear factor.

13.2 Factor $X^3 + 2X + 3$ over $\mathbb{Q}$, and prove that exactly one of the roots is a rational number. Then factor the polynomial over $\mathbb{C}$, approximating the coefficients of the factors.

13.3 Factor $X^8 - 2X^7 + X^6 + 14X^4 - 14X^3 + 49$ over $\mathbb{Q}$ and over $\mathbb{C}$, with approximations to the coefficients of the factors in the second case. How may we deduce that none of the roots is rational?

13.4 Show, by producing a counterexample, that the converse of Theorem 11.3 does not hold.

13.5 Suppose that a polynomial $p \in \mathbb{Q}[X]$ has four roots in $\mathbb{C}$, each with multiplicity 2. Prove that the polynomial must be the square of a polynomial q, where $q \in \mathbb{Q}[X]$. (Hint: Consider the minimal polynomial of one of the roots.) Then show that $X^8 - 2X^7 + X^6 + 14X^4 - 14X^3 + 49$ is an example of such a polynomial.

13.6 For the following $m_{\alpha,\mathbb{Q}}$ and β, find $m_{\beta,\mathbb{Q}}$ and show that it is irreducible.

(1) $m_{\alpha,\mathbb{Q}} = X^3 - X + 2$, $\beta = \alpha + 2$;

(2) $m_{\alpha,\mathbb{Q}} = X^5 + X^3 + X + 1$, $\beta = \frac{1}{2}\alpha - 3$;

(3) $m_{\alpha,\mathbb{Q}} = X^{10} - X^4 + 1$, $\beta = \alpha - 1$.

13.7 Let α be such that $m_{\alpha,\mathbb{Q}} = X^4 + X^2 + 1$. Let $\beta = \alpha^2$. Determine $m_{\beta,\mathbb{Q}}$, and deduce that β is an example of $\beta \in \mathbb{Q}(\alpha)$ such that $\mathbb{Q}(\beta) \neq \mathbb{Q}(\alpha)$.

B.

13.8* [Tower Law for Arbitrary Field Extensions] Following the proof of Theorem 12.9, prove that if M is a field containing a field L and L contains a field K, such that the dimension of M over L is finite and the dimension of L over K is finite, then $[M : K] = [M : L][L : K]$. Further prove that if the dimensions are arbitrary, then $[M : K]$ is infinite if and only if at least one of $[M : L]$ and $[L : K]$ is infinite.

13.9 Prove that if $p \in \mathbb{Q}[X]$ is an irreducible polynomial of degree $n > 1$ and $[K : \mathbb{Q}] = m$, where n and m are relatively prime, then p has no root in K.

13.10* Generalize Theorem 11.2 to arbitrary fields as follows: Prove that if $K \subset L$ are arbitrary fields and $0 \neq f \in K[X]$, then f has a unique representation

$$f = f(X) = c_n(X - \alpha_1)^{m_1}(X - \alpha_2)^{m_2} \cdots (X - \alpha_k)^{m_k} g(X),$$

where c_n is a constant in L, g is a monic polynomial in $L[X]$ with no roots in L, and the set $\{(\alpha_i, m_i)\}$ of pairs of roots $\alpha_i \in L$ with multiplicities m_i is uniquely determined by f (with respect to K and L). Use induction on the number of roots of f in L.

13.11* The derivative operation on polynomials can be defined in at least two ways.

One comes from the power and sum rules in calculus (though our definition will hold over an arbitrary field):

$$\left(\sum_{i=0}^{n} a_i X^i\right)' = \sum_{i=1}^{n} i a_i X^{i-1}.$$

Another, also valid over an arbitrary field, expresses the derivative of $f \in K[X]$ as the coefficient of h in the expression of $f(x+h)$ as a sum of $K[X]$-multiples of nonnegative powers of h. Using either definition, show that the derivative function $\cdot': K[X] \to K[X]$ is a homomorphism of additive groups.

13.12 Show that the kernel of the derivative operation on $K[X]$ for $K = \mathbb{Z}/p\mathbb{Z}$, the finite field with p elements, consists of more than the constant polynomials. For such a polynomial, where would the proof of Theorem 11.3 fail? We will see later that when K is such a field, the statement of Theorem 11.3 remains true. There exist still other fields for which the analogous statement is not true, however, such as the field in Exercise 35.5.

13.13 Prove Theorem 11.2 using Theorem 1.8.

13.14 Show that the statement of Corollary 11.4 does not hold for $p \in K[X]$ and K an arbitrary subfield of $\mathbb{C}$, by choosing $p = X^2 - e$, where e is any complex number that is not an algebraic number and $K = \mathbb{Q}(e)$, the set of all quotients of polynomials in e with coefficients in $\mathbb{Q}$. (Hint: First show that p is irreducible over K by showing that no square of any arithmetic combination of e can be equal to e. Then show that the square of an algebraic number must be an algebraic number.)

13.15 In the notation of the proof of Theorem 12.4, find formulas giving the coefficients of q in terms of a and b and the coefficients of p. You may wish first to experiment with *Maple* or *Mathematica*.

13.16 Find the minimal polynomial over $\mathbb{Q}$ for $\sqrt[4]{2} + 1$ by considering the polynomial over $\mathbb{Q}$ with roots $\sqrt[4]{2} + 1$, $i\sqrt[4]{2} + 1$, $-\sqrt[4]{2} + 1$, and $-i\sqrt[4]{2} + 1$.

14. Computation in Algebraic Number Fields: *Maple* and *Mathematica*

The package `AlgFields` implements a variety of functions for computation in fields of algebraic numbers. In this section, we consider functions for working with fields generated by an algebraic number over another field.

14.1. Declaring a Field

In order to work inside a field, we first declare the field to *Maple* or *Mathematica*. If the field is generated by a single algebraic number over the field of rational numbers, we provide two names: one for the field and one for the algebraic number that will generate the field. We also provide the minimal polynomial for the algebraic number. Since this polynomial does not uniquely determine the algebraic number, we may additionally specify a number that refers uniquely to a particular root of the minimal polynomial.

Using the function `FDeclareField`, we pass to the package the name of the field and then (in square brackets for *Maple*, in braces for *Mathematica*) the minimal polynomial of the algebraic number, using the name for the indeterminate. If we wish to specify the precise root number (corresponding, in *Maple*, to the number following `index=` in the `RootOf` function, or, in *Mathematica*, to the second argument of the `Root` function), we may also pass this number (again in brackets or braces).

We declare a field K to be $\mathbb{Q}(\alpha)$, where $m_\alpha = X^4 - 10X^2 + 20$.

```
>   FDeclareField(K,[alpha^4-10*alpha^2+20]);


----Details of field K ----

  Algebraic Numbers: [alpha]

  Dimension over Q: 4

  Minimal Polynomials: [

      alpha^4-10*alpha^2+20]

  Root Approximations: [

      1.6625077511098137144]

----
```

```
In[1]:= FDeclareField[K, {α^4 - 10α^2 + 20}]


----Details of field K ----

  Algebraic Numbers: {α}

  Dimension over Q: 4

  Minimal Polynomials: {

      20 - 10 * α^2 + α^4}

  Root Approximations: {

      -2.6899940478558293078}

----
```

Note that the field has been declared with respect to a particular root of $X^4 - 10X^2 + 20$, one that is a real number and is approximately 1.663 in *Maple* and is approximately -2.690 in *Mathematica*. In order to choose a different root, we first determine the complex approximations for each of the four roots. (We have deliberately chosen a polynomial with four real roots for this example, but `AlgFields` will certainly handle polynomials with imaginary roots as well.)

```
>   seq(evalf(RootOf(alpha^4-10*alpha^2+20,
>       index=i)),i=1..4);

        1.662507751, 2.689994048,
      -1.662507751, -2.689994048
```

```
In[2]:= Table[N[Root[α^4 − 10α^2 + 20, i]], {i, 1, 4}]

Out[2]= {−2.68999, −1.66251, 1.66251, 2.68999}
```

Then, we declare a field L to be $\mathbb{Q}(\beta)$, where $m_\beta = X^4 - 10X^2 + 20$ and β is the particular root that is approximately 2.690.

```
>   FDeclareField(L, [beta^4-10*beta^2+20],
>       [2]);


----Details of field L ----

  Algebraic Numbers: [beta]

  Dimension over Q: 4

  Minimal Polynomials: [
      beta^4-10*beta^2+20]

  Root Approximations: [

    2.6899940478558293078]


----
```

```
In[3]:= FDeclareField[L, {β^4 − 10β^2 + 20}, {4}]


----Details of field L ----

  Algebraic Numbers: {β}

  Dimension over Q: 4

  Minimal Polynomials: {20 - 10 * β^2 + β^4}

  Root Approximations: {

    2.6899940478558293078}

----
```

Now we may additionally want to work with fields generated by an algebraic number over a previously declared field. We do so with the function `FDeclareExtensionField`, to which we provide the name of the previously declared field as the second argument. We declare a field L2 to be $L(\gamma)$, where $m_{\gamma,L} = X^2 - \beta$.

```
>  FDeclareExtensionField(L2,L,[gam^2-beta]);
```

```
----Details of field L2 ----

  Algebraic Numbers: [beta, gam]

  Dimension over Q: 8

  Minimal Polynomials: [

      beta^4-10*beta^2+20,

      gam^2-beta]

  Root Approximations: [

      2.6899940478558293078,

      1.6401201321414932585]

----
```

```
In[4]:= FDeclareExtensionField[L2, L, {γ^2 − β}]
```

```
----Details of field L2----

  Algebraic Numbers: {β, γ}

  Dimension over Q: 8

  Minimal Polynomials: {

      20 − 10 * β^2 + β^4,

      −β + γ^2}

  Root Approximations: {

      2.6899940478558293078,

      −1.6401201321414932586}

----
```

(Here *Maple* and *Mathematica* chose different square roots of β.)

The `AlgFields` package permits the declaration of any field that may be reached by the successive generation by algebraic numbers, each number generating a field over the previous one, so long as the number of algebraic numbers used is finite (and the dimension is not *too* large!). We will take up the topic of multiply generated fields in more generality in the next chapter.

14.2. Reduced Forms

To bring an arithmetic combination without quotients into reduced form, we use the function `FSimplifyE`, and when quotients are required, we instead multiply by the inverse of the denominator, using the function `FInvert` to compute the inverse.

We reduce the expression $\alpha^{10} + \alpha^3 - 3\alpha^2 + 1/2$ in the field K. (Note that we continue to work with several objects throughout this section on computation.)

```
>   FSimplifyE(alpha^10+alpha^3-
>       3*alpha^2+1/2,K);
```
$$\alpha^3 + 4397\,\alpha^2 - \frac{23999}{2}$$

In[5]:= FSimplifyE[α^10 + α^3 − 3α^2 + 1/2, K]

Out[5]= $-\dfrac{23999}{2} + 4397\,\alpha^2 + \alpha^3$

We compute the reduced form of the inverse of α in K.

```
>   FInvert(alpha,K);
```
$$-\frac{1}{20}\,\alpha^3 + \frac{1}{2}\,\alpha$$

In[6]:= FInvert[α, K]

Out[6]= $\dfrac{\alpha}{2} - \dfrac{\alpha^3}{20}$

Now we put the expression $(\alpha^2 - 20\alpha - 5)/\alpha$ in reduced form.

```
>   FSimplifyE((alpha^2-20*alpha-5)*
>       FInvert(alpha,K),K);
```
$$\frac{1}{4}\,\alpha^3 - \frac{3}{2}\,\alpha - 20$$

In[7]:= FSimplifyE[(α^2 − 20α − 5) ∗ FInvert[α, K], K]

Out[7]= $-20 - \dfrac{3\,\alpha}{2} + \dfrac{\alpha^3}{4}$

To reduce the coefficients (without quotients) of a polynomial in one variable over a field, we use the function FSimplifyP.

```
>   FSimplifyP(x^2+alpha^10*x-20,x,K);
```
$$x^2 + (4400\,\alpha^2 - 12000)\,x - 20$$

In[8]:= FSimplifyP[x^2 + α^10x − 20, x, K]

Out[8]= $-20 + x^2 + x\,(-12000 + 4400\,\alpha^2)$

14.3. Factoring Polynomials over a Field

The function FFactor factors a polynomial in one variable over a declared field if the polynomial is expressed with coefficients without quotients.

```
>    FFactor(X^8+(-1+alpha)*X^7+
>        (20-9*alpha^2)*X^6+
>        (180+21*alpha+10*alpha^2-
>            10*alpha^3)*X^5+
>        (200-80*alpha^2)*X^4+
>        (4001+20*alpha-2000*alpha^2-
>            10*alpha^3)*X^3+
>        (-1+alpha)*X^2+alpha^2*X+
>        (200+alpha),K);
```

$$(X^3 - X^2 + X^2\alpha + \alpha^2 X + 200 + \alpha)$$

$$(X^5 + 20 X^3 - 10 X^3 \alpha^2 + 1)$$

$In[9]:=$ FFactor[X^8 + (−1 + α)X^7 +

$(20 − 9\alpha^2)$X^6 +

$(180 + 21\alpha + 10\alpha^2 −$

$10\alpha^3)$X^5 +

$(200 − 80\alpha^2)$X^4 +

$(4001 + 20\alpha − 2000\alpha^2$

$−10\alpha^3)$X^3 +

$(−1 + α)$X^2 + $α^2$X +

$(200 + α)$, K]

$Out[9]=$ $(200 − X^2 + X^3 + \alpha + X^2\alpha + X\alpha^2)$

$(1 + 20 X^3 + X^5 − 10 X^3 \alpha^2)$

For information on factoring polynomials over fields of algebraic numbers, see [38] and [40].

14.4. The Division Algorithm and Reduced Forms

The two functions FSimplifyE and FSimplifyP are valid whether or not the base field is $\mathbb{Q}$ or another declared field K, and they apply the Division Algorithm over K in their implementation. Hence the package provides functions for polynomial quotients and remainders in FPolynomialQuotient and FPolynomialRemainder.

```
>    FPolynomialQuotient(X^2+alpha*X+1,
>        X-2,X,K);
```

$$X + \alpha + 2$$

```
>    FPolynomialRemainder(X^2+alpha*X+1,
>        X-2,X,K);
```

$$5 + 2\alpha$$

$In[10]:=$ FPolynomialQuotient[X^2 + α ∗ X + 1,

X − 2, X, K]

$Out[10]=$ $2 + X + \alpha$

$In[11]:=$ FPolynomialRemainder[X^2 + α ∗ X + 1,

X − 2, X, K]

$Out[11]=$ $5 + 2\alpha$

14.5. The Euclidean Algorithm and Inverses

The function FInvert is also valid over declared fields and applies the Euclidean Algorithm in its implementation. Hence the package provides a function for the computation of polynomial greatest common divisors, in FPolynomialGCD and

`FPolynomialExtendedGCD`. The function `FPolynomialGCD` produces the GCD alone, while `FPolynomialExtendedGCD` produces the GCD and the coefficient polynomials of the $K[X]$-linear combination yielding the GCD.

```
> FPolynomialExtendedGCD(X^5+
>     (-alpha-alpha^2)*X^4+
>     alpha^3*X^3+5*X-5*alpha^2,
>     X^5+(-alpha-alpha^2)*X^4+
>     alpha^3*X^3+4*X-4*alpha^2,
>     X,K);
```
$$[X - \alpha^2, \ [1, \ -1]]$$

In[12]:= FPolynomialExtendedGCD[X^5+

$(-\alpha - \alpha^2)$X^4+

α^3X^3 + 5 X − 5 α^2,

X^5 + $(-\alpha - \alpha^2)$X^4+

α^3X^3 + 4 X − 4α^2,

X, K]

Out[12]= {X − α^2, {1, −1}}

14.6. Representing Algebraic Numbers and Finding Minimal Polynomials and Factors

Generally, there are two ways in which to specify an element δ of a field $K(\beta)$: either as an arithmetic combination of the generating algebraic number β of the field, or as a particular root of a polynomial. The package `AlgFields` allows us to move from one representation to the other.

When δ is expressed as an arithmetic combination of β, we may first eliminate any quotients appearing in the combination by using the `FInvert` function. Once an arithmetic combination for δ not containing quotients is obtained, the function `FMinPoly` returns a minimal polynomial for δ over any subfield of the enclosing field $K(\beta)$. Then, `FRootNumber` will determine which particular root of this minimal polynomial is δ.

Here we consider the minimal polynomial and particular root number of $\delta = \beta^3 + 7$ in the field L.

```
> FMinPoly(beta^3+7,X,L);
```
$$-9199 + 4228\, X - 106\, X^2 - 28\, X^3 + X^4$$

```
> FRootNumber(%,beta^3+7,L);
```
$$3$$

In[13]:= FMinPoly[β^3 + 7, X, L]

Out[13]= −9199 + 4228 X − 106 X^2 − 28 X^3 + X^4

In[14]:= FRootNumber[%, β^3 + 7, L]

Out[14]= 4

Going the other direction, assume that we have a polynomial for which δ is a root, and we have the particular number of the root. In order to express δ as an arithmetic combination of the generating algebraic number β, we use `FFindFactorRt` to factor the polynomial over the field and select an irreducible factor with δ as a root.[2] If δ does indeed lie in the declared field, then this factor will be of the form $aX + b$ for a and b arithmetic combinations of the field, and hence $\delta = -b/a$. Conversely, if the irreducible factor returned is linear, then δ clearly lies in the field. If the irreducible factor returned is not of degree 1, then δ is not in the declared field, but we have an irreducible polynomial over the field with δ as a root, which may be made monic in order to find the minimal polynomial for δ over the field.

First we find that an arithmetic combination in L for the "third" (*Maple*) or "fourth" (*Mathematica*) root of δ's minimal polynomial over $\mathbb{Q}$ is $\beta^3 + 7$; this root is δ itself.

```
>   FFindFactorRt(
>       X^4-28*X^3-106*X^2+4228*X-
>         9199,
>       RootOf(X^4-28*X^3-106*X^2+4228*X-
>         9199,index=3),L);
```

$$X - 7 - \beta^3$$

```
In[15]:= FFindFactorRt[
           X^4 − 28X^3 − 106X^2 + 4228X−
           9199,
           Root[X^4 − 28X^3 − 106X^2+
           4228X − 9199, 4], L]
Out[15]= −7 + X − β^3
```

Next we determine that the "first" (*Maple*) or "second" (*Mathematica*) root also lies in L.

```
>   FFindFactorRt(
>       X^4-28*X^3-106*X^2+4228*X-
>         9199,
>       RootOf(X^4-28*X^3-106*X^2+4228*X-
>         9199,index=1),L);
```

$$X - 7 - 20\beta + 3\beta^3$$

```
In[16]:= FFindFactorRt[
           X^4 − 28X^3 − 106X^2 + 4228X−
           9199,
           Root[X^4 − 28X^3 − 106X^2+
           4228X − 9199, 2], L]
Out[16]= −7 + X − 20 β + 3 β^3
```

We check that neither square root of 2 lies in L, however.

---

[2] A related function is `FFindFactor`, which assumes that δ is given not as a pair of a polynomial and the root number but instead as an arithmetic combination of the generating algebraic numbers. Then `FFindFactor` returns the factor of the polynomial which is zero at δ.

```
>   FFactor(X^2-2,L);
```

$$X^2 - 2$$

```
In[17]:= FFactor[X^2 - 2, L]
```

$$Out[17] = -2 + X^2$$

Since there is no linear factor in the factorization over L, no root of $X^2 - 2$ lies in L.

These procedures allow us to determine, for instance, if a field $\mathbb{Q}(\gamma_1)$ is a subfield of $\mathbb{Q}(\gamma_2)$. This problem is sometimes termed the *subfield immersion problem*. We need only determine whether or not γ_1 is an element of $\mathbb{Q}(\gamma_2)$. Moreover, by detecting whether or not γ_2 is an element of $\mathbb{Q}(\gamma_1)$, we may determine if two such fields are equal.

We first check that the algebraic number α generating K is contained in L. (Note that the symbolic computation systems chose different values for α.)

```
>   FFindFactorRt(X^4-10*X^2+20,
>       RootOf(X^4-10*X^2+20,index=1),L);
```

$$X + 3\beta - \frac{1}{2}\beta^3 \quad .$$

```
In[18]:= FFindFactorRt[X^4 - 10X^2 + 20,
         Root[X^4 - 10X^2 + 20, 1], L]
```

$$Out[18] = X + \beta$$

We find that α is in fact an element of L. Now we check that the algebraic number β generating L is contained in K.

```
>   FFindFactorRt(
>       X^4-28*X^3-106*X^2+4228*X-
>         9199,
>       RootOf(X^4-28*X^3-106*X^2+4228*X-
>         9199,index=2),K);
```

$$X - 7 - \alpha^3 \quad .$$

```
In[19]:= FFindFactorRt[
         X^4 - 28X^3 - 106X^2 + 4228X-
         9199,
         Root[X^4 - 28X^3 - 106X^2+
         4228X - 9199, 4], K]
```

$$Out[19] = -7 + X + \alpha^3$$

We find that β is in fact an element of K. Hence K and L are equal.

14.7. Reduced Forms over Subfields

Given δ in $\mathbb{Q}(\alpha)$, we may wish to express δ in reduced form in α over $\mathbb{Q}(\epsilon)$, where $\mathbb{Q}(\epsilon) \subsetneq \mathbb{Q}(\alpha)$, in order that the coefficients of α in the sum be allowed to take values in $\mathbb{Q}(\epsilon)$. To derive this reduced form in AlgFields, we use FMakeTower to declare a new field over K by adjoining ϵ and α successively, followed by FSimplifyE applied to δ over the newly declared field.

In our example, we begin with the field $L = \mathbb{Q}(\beta)$ from above. First we form the subfield generated over $\mathbb{Q}$ by $\epsilon = 4\beta^2 - 20$ ("`Lintermed`"), and then we form the field generated by ϵ and β ("`Lnew`"). Finally, we simplify an arithmetic combination $\beta^3 + 7$ into reduced form in β over $\mathbb{Q}(\epsilon)$ by simplifying the expression over `Lnew`.

```
>  FMakeTower(Lnew,Lintermed,L,
>  epsilon,4*beta^2-20);

----Details of field Lintermed ----

  Algebraic Numbers: [epsilon]

  Dimension over Q: 2

  Minimal Polynomials: [-80+epsilon^2]

  Root Approximations: [

    8.9442719099991587856]

----
```

```
In[20]:= FMakeTower[Lnew, Lintermed, L,
           ε, 4 * β^2 - 20]

----Details of field Lintermed ----

  Algebraic Numbers: {ε}

  Dimension over Q: 2

  Minimal Polynomials: {-80 + ε^2}

  Root Approximations: {

    8.944271909999158786}

----
```

```
----Details of field Lnew ----

  Algebraic Numbers: [epsilon, beta]

  Dimension over Q: 4

  Minimal Polynomials: [-80+epsilon^2,

    beta^2-5-1/4*epsilon]

  Root Approximations: [

    8.9442719099991587856,

    2.6899940478558293078]

----
```

```
----Details of field Lnew ----

  Algebraic Numbers: {ε, β}

  Dimension over Q: 4

  Minimal Polynomials: {-80 + ε^2,

    -5 + β^2 - ε/4}

  Root Approximations: {

    8.944271909999158786,

    2.6899940478558293086}

----
```

> FSimplifyE(beta^3+7,Lnew);

$$(5 + \frac{\varepsilon}{4})\,\beta + 7$$

`In[21]:=` FSimplifyE[β^3 + 7, Lnew]

`Out[21]=` $7 + \beta \left(5 + \frac{\epsilon}{4}\right)$

15. Exercise Set 2

A.

15.1 Factor $X^3 + 3X^2 + 22X - 27$ over $\mathbb{Q}(\alpha)$, where $m_{\alpha,\mathbb{Q}} = X^3 + X + 1$.

15.2 Factor $X^4 - X^2 - 6X - 2$ over $\mathbb{Q}(\alpha)$, where $m_{\alpha,\mathbb{Q}}$ is

(1) $X^2 - 3$;

(2) $X^2 - 9$;

(3) $X^2 - 27$;

(4) $X^2 + 2X - 2$;

(5) $X^2 - 5$.

Can you tell what is in common among the α for which the polynomial factors?

15.3 Let $\alpha = \sqrt[3]{-1}$ such that $\alpha \neq -1$. Perform the Division Algorithm over $\mathbb{Q}(\alpha)$ to divide $X^3 + \alpha X + 1$ by $X - \alpha^2$.

15.4 Factor $X^4 - 10X^2 + 5$ over $\mathbb{Q}(\alpha)$, where $m_{\alpha,\mathbb{Q}} = X^4 - 10X^2 + 5$. Deduce that four different roots of $X^4 - 10X^2 + 5$ can be expressed as arithmetic combinations of α.

15.5 Factor $X^4 + 3X + 1$ over $\mathbb{Q}(\alpha)$, where $m_{\alpha,\mathbb{Q}} = X^4 + 3X + 1$. Deduce that only one root of $X^4 + 3X + 1$ can be expressed as an arithmetic combination of α.

15.6 Let $m_{\alpha,\mathbb{Q}} = X^3 + 2X^2 + X + 1$. (a) Use the Euclidean Algorithm over $\mathbb{Q}(\alpha)$ to find a polynomial of maximal degree that is a common factor over $\mathbb{Q}(\alpha)$ of $X^2 + (-\alpha + \alpha^2)X + (2 + 2\alpha + 2\alpha^2)$ and $X^3 + \alpha X^2 + (3 - 6\alpha^2)X - 6\alpha$; (b) multiply this polynomial by an appropriate nonzero constant in $\mathbb{Q}(\alpha)$ to make it monic; (c) determine a common root for the two polynomials over $\mathbb{Q}(\alpha)$.

15.7 Let $m_{\alpha,\mathbb{Q}} = X^3 + 2X^2 + X + 1$. Factor the two polynomials $X^2 + (-\alpha + \alpha^2)X + (2 + 2\alpha + 2\alpha^2)$ and $X^3 + \alpha X^2 + (3 - 6\alpha^2)X - 6\alpha$ over $\mathbb{Q}(\alpha)$ to determine that they have a common root, and determine this common root in reduced form.

15.8 Let $m_{\alpha,\mathbb{Q}} = X^5 + X^2 + 1$. Find the minimal polynomial of $\alpha^2 + \alpha - 3$.

15.9 Find the minimal polynomial of $\sqrt[3]{2} + \sqrt[5]{5}$. (First declare a field containing both radicals.)

15.10 Let K be a subfield of $\mathbb{C}$. Prove that if $[K(\alpha) : K]$ is relatively prime to $[K(\beta) : K]$, then $[K(\alpha, \beta) : K] = [K(\alpha) : K][K(\beta) : K]$.

B.

15.11 Let $m_{\alpha,\mathbb{Q}} = X^6 + 3X^5 + 6X^4 + 7X^3 + 2X^2 - X + 1$ and $\beta = \alpha^2 + \alpha$. Check that $m_{\beta,\mathbb{Q}} = X^3 + 3X^2 - X + 1$. Determine $[\mathbb{Q}(\alpha) : \mathbb{Q}(\beta)]$, and rewrite $\alpha^4 - \alpha^2$ in reduced form in α over $\mathbb{Q}(\beta)$.

15.12 Let $m_{\alpha,\mathbb{Q}} = X^{10} + 4X^6 + 5X^5 + 4X^2 + 10X + 1$ and $\beta = \alpha^5 + 2\alpha$. Check that $m_{\beta,\mathbb{Q}} = X^2 + 5X + 1$. Determine $[\mathbb{Q}(\alpha) : \mathbb{Q}(\beta)]$, and rewrite $\alpha^7 + \alpha^5 + \alpha^3$ in reduced form in α over $\mathbb{Q}(\beta)$.

15.13 Let $m_{\alpha,\mathbb{Q}} = X^3 - 2X + 7$ and $\beta = \alpha^3 - 2\alpha$. Determine $m_{\beta,\mathbb{Q}}$ and $[\mathbb{Q}(\alpha) : \mathbb{Q}(\beta)]$, and rewrite $\alpha^6 + \alpha^5$ in reduced form in α over $\mathbb{Q}(\beta)$.

15.14 Let $m_{\alpha,\mathbb{Q}} = X^3 + 2X^2 - 5$ and $\beta = \alpha^2$. Verify that $m_{\beta,\mathbb{Q}} = X^3 - 4X^2 + 20X - 25$, determine $[\mathbb{Q}(\alpha) : \mathbb{Q}(\beta)]$, and rewrite α^4 in reduced form in α over $\mathbb{Q}(\beta)$.

15.15 Let $m_{\alpha,\mathbb{Q}} = X^5 - 15X^4 + 87X^3 - 250X^2 + 359X - 207$ with $\alpha \approx 5.619$. Determine if $\beta \approx -1.194$ with $m_{\beta,\mathbb{Q}} = X^5 + X^2 + 1$ is contained in $\mathbb{Q}(\alpha)$, and, if so, express β in reduced form in α.

15.16 Let $m_{\alpha,\mathbb{Q}} = X^9 + 3X^7 + 3X^5 + 2X^3 + X - 1$ and $\beta = -\alpha^3 - \alpha$. Rewrite $\alpha^8 + \alpha^3$ in reduced form over $\mathbb{Q}(\beta)$.

15.17 Let $m_{\alpha,\mathbb{Q}} = X^6 + 2X^4 + X^2 - 2$ and $\alpha \approx -0.834$. Find $m_{\alpha,\mathbb{Q}(\sqrt{2})}$.

15.18 Show that $\mathbb{Q}(\sqrt{d})$ is not a subset of $\mathbb{Q}(\alpha)$ when $\deg m_\alpha = 3$ and d is a nonsquare rational.

15.19 Let K be a subfield of $\mathbb{C}$. Prove that if $[K(\alpha) : K]$ is odd, then $K(\alpha) = K(\alpha^2)$. Give a counterexample to the statement for $[K(\alpha) : K]$ even.

Multiply Generated Fields

We now extend our investigation to fields generated over a field K by more than one algebraic number, or *multiply* generated fields. We saw a hint of such fields in section 12.3, where we considered a field $K(\alpha)$ generated by an algebraic number α over a field $K(\beta)$, itself generated over K by an algebraic number β. In this chapter, we consider several questions concerning fields generated over K by more than one algebraic number:

- whether such a field may be generated over K by a single algebraic number;
- whether such a field is of finite dimension over K; and
- how to specify the structure of such fields via an isomorphism to a quotient ring of a polynomial ring, as we did in section 8.3.

We also examine an important class of fields generated by several algebraic numbers: those that are generated by all of the roots of a polynomial. We then study how to determine isomorphisms from one multiply generated field to another, particularly when the fields are splitting fields. At the end of the chapter, we consider the results of this chapter in the general field-theoretic setting of simple, finite, and algebraic extensions.

16. Fields Generated by Several Algebraic Numbers

Definition 16.1. Let K be a subfield of $\mathbb{C}$, and let α and β be two algebraic numbers. We denote by $K(\alpha, \beta)$ the identical fields $K(\alpha)(\beta) = K(\beta)(\alpha)$, which we call the *field generated by α and β over K*.

In support of the preceding definition, note that we may view the set S of all arithmetic combinations of α, β, and elements of K in several ways. On one hand, any such arithmetic combination is an arithmetic combination of β <u>over the field $K(\alpha)$</u>, so that this set S must be the same as the field $K(\alpha)(\beta)$, that is, the field generated by β over the field $K(\alpha)$. On the other hand, the set S is the set of arithmetic combinations of α with elements of $K(\beta)$, or $K(\beta)(\alpha)$. Since the order is therefore irrelevant, we separate the numbers by a comma and place them both in the parenthesized expression, as in $K(\alpha, \beta)$ or $K(\beta, \alpha)$, in order to stress that the result is the same.

16.1. Generation by Two Algebraic Numbers Is Generation by One

Given K a subfield of $\mathbb{C}$ and α and β two algebraic numbers, a first question is whether $K(\alpha, \beta)$ is of finite dimension when considered as a vector space over K. Essentially, we already know that this question is answered affirmatively.

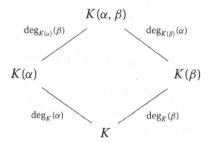

Viewing the field with the help of the subspace diagram we use Exercise 13.8 and Theorem 8.6 to conclude that

$$[K(\alpha, \beta) : K] = [K(\alpha, \beta) : K(\alpha)][K(\alpha) : K] \le [K(\beta) : K][K(\alpha) : K],$$

since $m_{\beta, K(\alpha)} \mid m_{\beta, K}$.

Now that we know that the dimension of the extension is finite over the field K, we may further ask whether the field may be generated by a single algebraic number γ over K. If so, we will say that the algebraic number γ is a *primitive element* for the field extension $K(\alpha, \beta)/K$, and we will call the field extension a *simple field extension* of K. The following theorem guarantees that field extensions generated by two algebraic numbers are simple extensions.

Theorem 16.2 (Generation by Two Algebraic Numbers Is Generation by One). *Let K be a subfield of $\mathbb{C}$, and let α and β be algebraic numbers. Then the field $K(\alpha, \beta)$ is a simple extension of K, that is, there exists an algebraic number γ such that $K(\alpha, \beta) = K(\gamma)$.*

Proof. Our goal will be to show that except for finitely many rational numbers t, elements of the form $t\alpha + \beta$ are algebraic numbers that generate $K(\alpha, \beta)$ over K.

Let $\alpha_1 = \alpha$ and $\alpha_2, \ldots, \alpha_k$ be the roots of m_α in $\mathbb{C}$, and let $\beta_1 = \beta$ and $\beta_2, \ldots, \beta_l$ be the roots of m_β in $\mathbb{C}$.

Consider $s \in \mathbb{Q}$ such that $s \neq (\beta_i - \beta_1)/(\alpha_j - \alpha_1)$ for all $i \geq 2$, $j \geq 2$. Let L be the field generated over K by $\beta_1 - s\alpha_1$, and set p to be the polynomial in $L[X]$ defined by $p(X) = m_\beta(sX - s\alpha_1 + \beta_1)$.

Because of the condition on s, the only α_i that is a root of p is α_1. Now because the roots of a minimal polynomial are distinct (Theorem 11.3), the greatest common divisor of m_α and p in $\mathbb{C}[X]$ is the linear factor $X - \alpha_1$. However, the Euclidean Algorithm (Theorem 1.7) determines the greatest common divisor of two polynomials without extending the field of coefficients (see Exercise 5.4). Hence the greatest common divisor $X - \alpha_1$ lies in $L[X]$, whence $\alpha_1 \in L$.

But then $\beta_1 = (\beta_1 - s\alpha_1) + s\alpha_1$ is in L, in which case L contains both $\alpha_1 = \alpha$ and $\beta_1 = \beta$. Since L is a subfield of $K(\alpha, \beta)$, we must have that $L = K(\alpha, \beta)$. If we let $\gamma = \beta - s\alpha$, then, as an element of a field $\mathbb{Q}(\alpha, \beta)$ generated over a field of algebraic numbers $\mathbb{Q}(\alpha)$ by an algebraic number β, γ is algebraic (Theorem 12.12), and we are done. $\qquad \square$

The fact that in the proof of Theorem 16.2, the primitive element γ takes the particular form of a K-linear combination of α and β, is not important. Over subfields of $\mathbb{C}$, any sufficiently random number in $K(\alpha, \beta)$ will be a primitive element for $K(\alpha, \beta)/K$. To determine if a particular element $\gamma \in K(\alpha, \beta)$ is a primitive element for the extension, we can find its minimal polynomial over K and check if its degree is equal to the dimension $[K(\alpha, \beta) : K]$. In sections 21.7 and 21.8, we describe computational tools for these operations, which we use to work out the following example.

Example 16.3. We find a primitive element for the field $L = \mathbb{Q}(\alpha, \beta)$, with $\alpha = \sqrt[3]{2} \approx 1.26$ and $\beta = \sqrt{3} \approx 1.73$. Since $m_\alpha(X) = X^3 - 2$ and $m_{\beta, \mathbb{Q}(\alpha)}(X) = X^2 - 3$, $[L : \mathbb{Q}] = 6$.

Let $\gamma = \alpha + \beta$. We verify that the minimal polynomial of γ over L is

$$X^6 - 9X^4 - 4X^3 + 27X^2 - 36X - 23,$$

and therefore $[\mathbb{Q}(\gamma) : \mathbb{Q}] = 6$. Since $\mathbb{Q} \subset \mathbb{Q}(\gamma) \subset L$, we deduce that $L = \mathbb{Q}(\gamma)$.

By factoring $X^3 - 2$ and $X^2 - 3$ over $\mathbb{Q}(\gamma)$ and looking for linear factors, we may then express α and β in terms of γ:

$$\alpha = \frac{1}{102} \left(-4\gamma^5 - \gamma^4 + 40\gamma^3 + 26\gamma^2 - 76\gamma + 91\right)$$

$$\beta = \frac{1}{102} \left(4\gamma^5 + \gamma^4 - 40\gamma^3 - 26\gamma^2 + 178\gamma - 91\right)$$

16.2. From Multiply Generated Extensions to Multivariate Polynomial Rings

In section 8.3, we determined that each field $K(\alpha)$ generated over K by an algebraic number α with minimal polynomial $m_{\alpha,K}$ is isomorphic to the quotient $K[X]/(m_{\alpha,K}(X))$ of the polynomial ring $K[X]$. By transitivity of isomorphism, all of these fields $K(\alpha')$ for conjugates α' of α are isomorphic: for every pair of conjugates α_i, α_j with minimal polynomial $m_{\alpha,K}$, there exists an isomorphism

$$\epsilon_{i,j} \colon K(\alpha_i) \to K[X]/(m_{\alpha,K}(X)) \to K(\alpha_j). \tag{16.1}$$

As we will see later, we can even arrange that $\epsilon_{i,j}(\alpha_i) = \alpha_j$.

We wish to take up the question of whether this result can be extended to fields generated by finitely many algebraic numbers. We will find that the result does not generalize directly, and we will study situations in which counterexamples occur. In doing so, we will lay the groundwork for an understanding of an important class of fields generated by several algebraic numbers, *splitting fields*.

Now we might continue our study of multiply generated fields in two ways. On one hand, we could invoke Theorem 16.2 inductively (Exercise 20.4) to consider a multiply generated field $K(\alpha_1, \alpha_2, \ldots, \alpha_n)$ as a simple extension $K(\delta)$ of K. On the other hand, we could work to determine a *formal object* for $K(\alpha_1, \alpha_2, \ldots, \alpha_n)$, as we did in Theorem 8.12 for $K(\alpha)$. In this context, we would expect a formal object to be some quotient of a polynomial ring (in n variables) that is isomorphic to $K(\alpha_1, \alpha_2, \ldots, \alpha_n)$. For the purposes of determining whether the result above generalizes, we choose the latter route, since the size and complexity of $m_{\delta,K}$ often obscures – at least computationally, if not theoretically – the relationships among the α_i.

To produce a formal object for $L = K(\alpha_1, \alpha_2, \ldots, \alpha_n)$, we begin by considering polynomial rings in n variables. (From here on we assume that $n > 1$.)

Definition 16.4. Let K be a field. The *polynomial ring* $K[X_1, X_2, \ldots, X_n]$ *in n variables over K* is the set of finite formal sums

$$\left\{ \sum a_{i_1, i_2, \ldots, i_n} X_1^{i_1} X_2^{i_2} \cdots X_n^{i_n} \;\middle|\; i_1, i_2, \ldots, i_n \in \mathbb{N} \cup \{0\}, \; a_{i_1, i_2, \ldots, i_n} \in K \right\}.$$

The elements of $K[X_1, \ldots, X_n]$ are called *multivariate polynomials over K*, and the addition and multiplication operations are the natural ones in this commutative ring. When we wish to indicate the variables in a polynomial $p \in K[X_1, \ldots, X_n]$, we write $p(X_1, \ldots, X_n)$, and we denote the evaluation of p at quantities $X_i = c_i$ by $p(c_1, \ldots, c_n)$.

It is an exercise (20.1) to define polynomial rings $D[X_1, \ldots, X_n]$ in n variables over an integral domain D and to show that these rings are themselves integral domains.

Frequently it is useful to consider polynomials $p \in K[X_1, \ldots, X_n]$ as polynomials in one distinguished variable X_n, with coefficients that are polynomials in $K[X_1, \ldots, X_{n-1}]$. In other words, we view $K[X_1, \ldots, X_n]$ as $K[X_1, \ldots, K_{n-1}][X_n]$: the ring of polynomials in X_n with coefficients in $K[X_1, \ldots, X_{n-1}]$. For example,

$$p(X_1, X_2) = 1 + X_1 X_2 + X_1 X_2^2 + X_1^2 X_2 \in K[X_1, X_2]$$

may be written

$$p(X_1, X_2) = (1) + (X_1 + X_1^2) X_2 + (X_1) X_2^2 \in K[X_1][X_2].$$

We denote by $\deg_{X_n} p$ the degree of p considered as a polynomial in $K[X_1, \ldots, X_{n-1}][X_n]$. This degree is the highest power of X_n occurring in p.

Now the ring $K[X_1, \ldots, X_n]$ is not a principal ideal domain; it is not true that each ideal I is the set of polynomial multiples $\{fp \mid p \in K[X_1, \ldots, X_n]\}$ of a single polynomial $f \in K[X_1, \ldots, X_n]$. Consequently, we turn to a more general form for expressing ideals.

We write

$$(f_1, f_2, \ldots, f_i) = \left\{ f_1 p_1 + f_2 p_2 + \cdots + f_i p_i \;\middle|\; p_1, p_2, \ldots, p_i \in K[X_1, \ldots, X_n] \right\}$$

for the set of all sums of polynomial multiples of the polynomials $f_1, f_2, \ldots, f_i$. We say that $(f_1, f_2, \ldots, f_i)$ is the *ideal generated by the polynomials* $f_1, f_2, \ldots, f_i$. It is not difficult to check that $(f_1, f_2, \ldots, f_i)$ is in fact an ideal, and we will seek to understand ideals of $K[X_1, \ldots, X_n]$ by expressing them in this form where possible.

16.3. Fields Generated by a Finite Number of Algebraic Numbers Are Quotients of Polynomial Rings

With the notion of multivariate polynomials in hand, we return to multiply generated fields $K(\alpha_1, \ldots, \alpha_n)$.

What should a formal object for $K(\alpha_1, \ldots, \alpha_n)$ be? For fields generated by a single algebraic number, we took the polynomial ring $K[X]$ modulo the kernel of the *evaluation homomorphism* $\varphi \colon K[X] \to K(\alpha)$ given by $\varphi(f) = f(\alpha)$, and we discovered that the kernel was the ideal $(m_{\alpha, K}(X))$. A first guess for $K(\alpha_1, \ldots, \alpha_n)$ would be that we want to take the polynomial ring $K[X_1, \ldots, X_n]$ modulo the kernel of the evaluation homomorphism

$$\varphi \colon K[X_1, \ldots, X_n] \to K(\alpha_1, \ldots, \alpha_n), \qquad \varphi(f) = f(\alpha_1, \ldots, \alpha_n),$$

and we expect to discover that $\ker \varphi = (m_{\alpha_1, K}(X_1), \ldots, m_{\alpha_n, K}(X_n))$.

To see that this first guess is not quite right, look at the case $\mathbb{Q}(\alpha_1, \alpha_2)$ with $\alpha_1 = \sqrt{2}$ and $\alpha_2 = -\sqrt{2}$. Consider the evaluation homomorphism

$$\varphi \colon \mathbb{Q}[X_1, X_2] \to \mathbb{Q}(\sqrt{2}), \qquad \varphi(f) = f(\sqrt{2}, -\sqrt{2}).$$

Certainly $m_{\alpha_1}(X_1) = X_1^2 - 2$ and $m_{\alpha_2}(X_2) = X_2^2 - 2$ lie in the kernel, leading us to suspect that $\ker \varphi = (X_1^2 - 2, X_2^2 - 2)$. However, the polynomial $X_1 + X_2$ also lies in the kernel, and it is not an element of the ideal $(X_1^2 - 2, X_2^2 - 2)$, as the following argument shows:

Suppose

$$X_1 + X_2 = (X_1^2 - 2)\, p_1(X_1, X_2) + (X_2^2 - 2)\, p_2(X_1, X_2).$$

Substitute X for X_1 and X_2. Then we have

$$2X = (X^2 - 2)(p_1(X) + p_2(X)),$$

which is impossible: the right-hand side must either be zero or have degree at least 2.

In hindsight, the fact that $(X_1^2 - 2, X_2^2 - 2)$ is not the kernel is not surprising. While X_1 and X_2 are not related, $\sqrt{2}$ and $-\sqrt{2}$ are, and this relation cannot be captured by taking polynomial multiples of the two minimal polynomials. In order to construct the kernel, we should begin with the minimal polynomial for $\alpha_1 \colon m_{\alpha_1}(X_1) = X_1^2 - 2$. Then we must use a polynomial which, by lying in the kernel of the evaluation homomorphism, expresses the relationship $\alpha_2 = -\alpha_1$, and this polynomial is $X_2 + X_1$. In fact, we can derive such a polynomial from a minimal polynomial. The minimal polynomial we use, however, is not $m_{\alpha_2}(X_2)$, but $m_{\alpha_2, \mathbb{Q}(\alpha_1)}(X_2)$!

We now describe how to construct the ideal generators f_i for $\ker \varphi$. Let $K_0 = K$ and $K_i = K(\alpha_1, \ldots, \alpha_i)$ for $i = 1, \ldots, n$. Set $d_i = \deg_{K_{i-1}}(\alpha_i)$ and consider the minimal polynomial

$$m_{\alpha_i, K_{i-1}}(X) = X^{d_i} + \sum_{j=0}^{d_i - 1} c_j X^j$$

of α_i over K_{i-1}. The coefficients c_j of this polynomial are elements of K_{i-1}, and therefore they may be expressed in reduced form in $\alpha_1, \ldots, \alpha_{i-1}$ over K. Exercise 20.5 makes this notion of a reduced form precise: each c_j may be expressed as a sum of K-multiples of products of powers of $\alpha_1, \ldots, \alpha_{i-1}$, where no power of an α_k exceeds $d_k - 1$ for $k = 1, \ldots, i - 1$.

Now, for each coefficient c_j expressed as such a sum, replace each α_k, $k = 1, \ldots, i - 1$, by the variable X_k. The minimal polynomial $m_{\alpha_i, K_{i-1}}(X)$ with these replacements becomes a polynomial in the variables X and $X_1, X_2, \ldots, X_{i-1}$, and we denote it by $f_i(X_1, X_2, \ldots, X_{i-1}, X)$. Observe that running our process in reverse,

$$f_i(\alpha_1, \ldots, \alpha_{i-1}, X) = m_{\alpha_i, K_{i-1}}(X), \tag{16.2}$$

or, substituting X_i for X, $f_i(\alpha_1, \ldots, \alpha_{i-1}, X_i) = m_{\alpha_i, K_{i-1}}(X_i)$.

We are now ready to establish our formal object for multiply generated fields.

Theorem 16.5 (Structure Theorem for Multiply Generated Fields). *Let K be a subfield of $\mathbb{C}$ and $\alpha_1, \ldots, \alpha_n$ algebraic numbers. Define the multivariate polynomials $f_i \in K[X_1, \ldots, X_n]$ as above. Then*

$$K(\alpha_1, \ldots, \alpha_n) \cong K[X_1, \ldots, X_n]/I$$

where I is the ideal

$$I = (f_1(X_1), f_2(X_1, X_2), \ldots, f_n(X_1, \ldots, X_n)).$$

Proof. Let $\varphi : K[X_1, \ldots, X_n] \to K(\alpha_1, \ldots, \alpha_n)$ be the natural *evaluation homomorphism* given by

$$\varphi(f(X_1, \ldots, X_n)) = f(\alpha_1, \ldots, \alpha_n).$$

We will show that $\ker \varphi$ is the ideal $I = (f_1, \ldots, f_n)$, and then the result will follow by the First Isomorphism Theorem for Rings (3.1).

Let $K_0 = K$. For $i = 1, \ldots, n$, let $K_i = K(\alpha_1, \ldots, \alpha_i)$ and $d_i = \deg_{K_{i-1}}(\alpha_i)$. Let V be the set of K-linear combinations of the monomials in the set

$$S = \{X_1^{a_1} X_2^{a_2} \cdots X_n^{a_n}\}_{0 \leq a_i < d_i}.$$

Clearly V is the subset of $K[X_1, \ldots, X_n]$ consisting of all polynomials such that, for all i, the highest power of X_i occurring is less than d_i.

We claim that φ is an onto homomorphism and that, moreover, $\varphi(V) = K_n$; in other words, every element $\beta \in K_n$ is the image of some $v \in V$ under φ. Now by Exercise 20.5, each element $\beta \in K_n$ may be expressed in reduced form over K: a sum of products of powers of the α_i, with each α_i occurring to powers no higher than $d_i - 1$. Replacing each α_i with X_i, we have a polynomial $v_\beta \in K[X_1, \ldots, X_n]$ in V. Moreover, by construction, $\varphi(v_\beta) = \beta$. Hence $\varphi(V) = K_n$.

A repeated application of the Tower Law (Exercise 13.8) shows that

$$d_1 d_2 \cdots d_n = [K_1 : K_0][K_2 : K_1] \cdots [K_n : K_{n-1}] = [K_n : K]. \tag{16.3}$$

Observe that $d_1 d_2 \cdots d_n$ is also the number of elements of the basis S of V. Hence V and K_n are vector spaces of the same dimension over K, and φ, as an onto ring homomorphism, is an onto linear transformation from V to K_n. We conclude that φ is also one-to-one, so that $V \cap \ker \varphi = \{0\}$.

We turn next to I. First, we claim that $I \subset \ker \varphi$. By definition, I consists of sums of polynomial multiples of the f_i. By equation (16.2), for each generator f_i of I,

$$\varphi(f_i) = f_i(\alpha_1, \ldots, \alpha_i) = m_{\alpha_i, K_{i-1}}(\alpha_i) = 0.$$

Hence by the homomorphism property, $\varphi(f) = 0$ for all $f \in I$.

Now we show that $V + I = K[X_1, \ldots, X_n]$, that is, that every polynomial $p \in K[X_1, \ldots, X_n]$ may be expressed as the sum of some polynomial $p_V \in V$ and some polynomial $p_I \in I$. We proceed by induction on i, showing that $K[X_1, \ldots, X_i] \subset V + I$. Our result is then the case $i = n$.

For the base case $i = 1$, let $p \in K[X_1]$ be arbitrary. Using the Division Algorithm (Theorem 1.5), write

$$p = f_1(X_1)q(X_1) + r(X_1),$$

with $r(X_1) = 0$ or $\deg r < \deg f_1 = d_1$. Set $p_I = f_1 q \in I$ and $p_V = r$, which by virtue of its degree lies in V. Then $p = p_V + p_I \in V + I$, as desired.

For the induction step, assume that $K[X_1, \ldots, X_i] \subset V + I$ and consider an arbitrary $p \in K[X_1, \ldots, X_{i+1}]$. Consider $K[X_1, \ldots, X_{i+1}]$ as $K[X_1, \ldots, X_i][X_{i+1}]$, the ring of polynomials in X_{i+1} over the integral domain $K[X_1, \ldots, X_i]$. Using the Division Algorithm for Integral Domains (Exercise 5.14), divide p by the monic polynomial f_{i+1}:

$$p = f_{i+1}(X_1, \ldots, X_{i+1})q(X_1, \ldots, X_{i+1}) + r(X_1, \ldots, X_{i+1}),$$

where $r = 0$ or $\deg_{X_{i+1}} r < \deg_{X_{i+1}} f_{i+1} = d_{i+1}$.

Now write

$$r(X_1, \ldots, X_{i+1}) = \sum_{j=0}^{d_{i+1}-1} r_j(X_1, \ldots, X_i) X_{i+1}^j.$$

Since each $r_j \in K[X_1, \ldots, X_i]$, by induction $r_j = (r_j)_V + (r_j)_I$ for $(r_j)_V \in V$ and $(r_j)_I \in I$. Then

$$r = \left(\sum_{j=0}^{d_{i+1}-1} (r_j)_V X_{i+1}^j \right) + \left(\sum_{j=0}^{d_{i+1}-1} (r_j)_I X_{i+1}^j \right), \tag{16.4}$$

and we let r_V and r_I be the two parenthesized expressions in equation (16.4), respectively. Checking degrees, we see that $r_V \in V$, while r_I, as a sum of polynomial multiples of elements of I, lies in I.

Now let $p_I = f_{i+1}q + r_I \in I$ and $p_V = r_V \in V$. By construction, $p \in V + I$, as desired.

We are finally ready to show that $\ker \varphi \subset I$ and hence that $\ker \varphi = I$. Let $p \in \ker \varphi$. Since $K[X_1, \ldots, X_n] = V + I$, $p = p_V + p_I$ with $p_V \in V$, $p_I \in I$. Because $I \subset \ker \varphi$, $\varphi(p_I) = 0$. On the other hand, $\varphi(p) = 0$ implies that

$$\varphi(p_V) = \varphi(p - p_I) = \varphi(p) - \varphi(p_I) = 0.$$

But $V \cap \ker \varphi = \{0\}$. Hence $p_V = 0$. Therefore $p = p_I \in I$ for each $p \in \ker \varphi$, and we are done. □

16.4. Splitting Fields

We now come to one of the most important classes of fields in Galois theory, that of splitting fields.

Definition 16.6 (Splitting Field). Let K be a subfield of $\mathbb{C}$ and $p \in K[X]$ a nonzero polynomial. The *splitting field over K of p* is the field $L = K(\alpha_1, \alpha_2, \ldots, \alpha_n)$, where $\{\alpha_i\}$ is the set of roots of p. We also say that L/K is a *splitting field extension*.

Splitting fields are so named because the polynomial p splits, or *splits completely*, under factorization: p factors into linear terms over the splitting field.

Here we bound the size of a splitting field of an irreducible polynomial.

Theorem 16.7. *Let K be a subfield of $\mathbb{C}$, $\{\alpha_1, \ldots, \alpha_n\}$ the set of roots of an irreducible polynomial over K, and $L = K(\alpha_1, \ldots, \alpha_n)$. Then the dimension $[L : K]$ satisfies $n \le [L : K] \le n!$.*

Proof. Since the number of roots of a minimal polynomial is equal to the degree of the polynomial (Theorem 11.3), and the subfield $K(\alpha_1)$ has dimension over K equal to the degree of this polynomial (Theorem 8.6), the dimension of the splitting field over K must be at least n.

On the other hand, by equation (16.3) and using the calculation of the dimensions of fields from Theorem 8.6, we find that the degree $[K(\alpha_1, \ldots, \alpha_n) : K]$ is equal to

$$(\deg m_{\alpha_1, K})(\deg m_{\alpha_2, K(\alpha_1)}) \cdots (\deg m_{\alpha_n, K(\alpha_1, \ldots, \alpha_{n-1})}).$$

Now the minimal polynomial for α_i over $K(\alpha_1, \ldots, \alpha_{i-1})$ cannot have degree greater than $n - i + 1$, since $(X - \alpha_1) \cdots (X - \alpha_{i-1})$ is a factor of $m_{\alpha_i, K}$ over $K(\alpha_1, \ldots, \alpha_{i-1})$; hence the product of the degrees can be no more than $n!$. $\square$

The bounds in the preceding theorem are, in fact, sharp, though a proof that for every n there are polynomials with splitting field of degree $n!$ is beyond the scope of this text. See [1, §14.8] or [6, §4.16, Ex. 5] for a sketch.

17. Characterizing Isomorphisms between Fields: Three Cubic Examples

In this section, we use the Structure Theorem for Multiply Generated Fields (16.5) to explore the question of the existence of isomorphisms between fields generated by more than one algebraic number over a common subfield K.

Suppose, for instance, that we have $K_n = K(\alpha_1, \ldots, \alpha_n)$ and a field $L \subset \mathbb{C}$ that is a field extension of K, and we wish to determine if there exists an isomorphism from K_n to L that leaves each element of K fixed. Let I be the ideal of the Structure Theorem. Since K_n is isomorphic to the quotient of $K[X_1, \ldots, X_n]$ by I, K_n is isomorphic to L if and only if $K[X_1, \ldots, X_n]/I$ is isomorphic to L.

Now every homomorphism $\varphi: K[X_1, \ldots, X_n]/I \to L$ is uniquely determined by the images $\varphi(X_i + I)$ of $X_i + I$ for $1 \le i \le n$. Our plan is to investigate what relationships among these images $\beta_i = \varphi(X_i + I) \in L$ must hold because φ is a homomorphism, and what else must hold if moreover φ is to be an isomorphism.

Assume then that φ is a homomorphism from $K[X_1, \ldots, X_n]/I$ to L. Observe first that because φ is a homomorphism and $\varphi(0 + I) = 0$, it follows that for each ideal generator $f_i(X_1, \ldots, X_i)$ of $I = (f_1, \ldots, f_n)$,

$$\varphi(f_i) = f_i(\beta_1, \ldots, \beta_i) = 0. \tag{17.1}$$

(Why?) Now let's explore what this means in the two cases $i = 1$ and $i = 2$.

In the case $i = 1$, we have $f_1 = m_{\alpha_1, K}(X_1)$. By equation (17.1), β_1 is a root of $f_1(X_1)$, and so β_1 has the same minimal polynomial as α_1 over K. In other words, α_1 and β_1 are conjugates over K. So far, so good.

In the case $i = 2$, by equation (16.2), $f_2(\alpha_1, X_2) = m_{\alpha_2, K(\alpha_1)}(X_2)$. However, by equation (17.1), $f_2(\beta_1, \beta_2) = 0$ in L. Hence β_2 is a root of the polynomial $f_2(\beta_1, X_2)$. Look closely: we need to pay attention to the difference between $f_2(\alpha_1, X_2)$ and $f_2(\beta_1, X_2)$. Observe that since the substitution $X_1 = \alpha_1$ makes $f_2(X_1, X_2)$ into $m_{\alpha_2, K(\alpha_1)}(X_2)$, then when we make the substitution $X_1 = \beta_1$ in $f_2(X_1, X_2)$,

we are replacing α_1 by β_1 in the coefficients of $m_{\alpha_2, K(\alpha_1)}$.

In other words, we are applying an isomorphism $\epsilon_1: K(\alpha_1) \to K(\beta_1)$, with $\epsilon_1(\alpha_1) = \beta_1$, to the coefficients of the polynomial $m_{\alpha_2, K(\alpha_1)}$:

$$f_2(\beta_1, X_2) = \epsilon_1(m_{\alpha_2, K(\alpha_1)}(X_2)).$$

Then, since $f_2(\beta_1, \beta_2) = 0$, β_2 is a root of an *alteration* of the polynomial $m_{\alpha_2, K(\alpha_1)}(X_2)$. In short, while α_1 and β_1 are conjugates over K, α_2 and β_2 are roots of similar *but possibly distinct* polynomials: $m_{\alpha_2, K(\alpha_1)}(X)$ for the former, and $\epsilon_1(m_{\alpha_2, K(\alpha_1)}(X))$ for the latter.

Here we reach the theme of this section, which bridges the Structure Theorem in the last section and the many results on isomorphisms between fields in the next. The existence of an isomorphism ϵ from a multiply generated field $K_n = K(\alpha_1, \ldots, \alpha_n)$ to a field extension $L \supset K$ essentially depends on

- finding elements $\beta_i \in L$, $1 \le i \le n$,
- where each β_i is the root of a polynomial $F_i \in K(\beta_1, \ldots, \beta_{i-1})[X]$, and moreover

- each F_i is obtained from $m_{\alpha_i, K(\alpha_1, \dots, \alpha_{i-1})}$ by applying an isomorphism

$$\epsilon_{i-1} \colon K(\alpha_1, \dots, \alpha_{i-1}) \to K(\beta_1, \dots, \beta_{i-1})$$

satisfying $\epsilon_{i-1}(\alpha_k) = \beta_k$, $1 \le k \le i - 1$.

If these conditions are satisfied, there exists a monomorphism $\epsilon = \epsilon_n \colon K_n \to L$ with $\epsilon(\alpha_i) = \beta_i$ for $1 \le i \le n$. (For simplicity, set $\epsilon_0 \colon K \to K$ to be the identity. Later we handle the question of building a monomorphism from K_n to L from a nonidentity isomorphism ϵ_0 from K to K.)

We will prove in the next section, with Theorem 18.3, that these conditions, together with the requirement that L is generated over K by the β_i, are *necessary and sufficient* for the existence of an isomorphism from K_n to L.

Before doing so, however, we examine these conditions closely in three examples. We will return to these examples in Chapter 5, as Examples 24.14–24.16. These examples involve the special case in which L is equal to K_n, so that while we know that at least one isomorphism exists from K_n to L (where $\beta_i = \alpha_i$ for each i), we wish instead to find *all* such isomorphisms. Furthermore, our isomorphisms will leave the common base field K elementwise fixed.

Example 17.1. (Compare Example 24.14.) Let α_1, α_2, and α_3 be the three roots of $m_{\alpha_i, \mathbb{Q}} = X^3 + 3X + 1$, where

$$\alpha_1 \approx -0.322, \quad \alpha_2 \approx 0.161 - 1.754i, \quad \text{and } \alpha_3 \approx 0.161 + 1.754i.$$

Let $K = K_0 = \mathbb{Q}$ and $K_i = \mathbb{Q}(\alpha_1, \dots, \alpha_i)$ for $i = 1, 2, 3$.

We first calculate the I of the Structure Theorem (16.5) such that $\mathbb{Q}[X_1, X_2, X_3]/I \cong K_3$. To do so, we need the three minimal polynomials $m_{\alpha_i, K_{i-1}}(X_i)$. Computing these, for instance, with either *Maple* or *Mathematica* (see section 21), we find

$$m_{\alpha_1, K_0}(X_1) = X_1^3 + 3X_1 + 1;$$

$$m_{\alpha_2, K_1}(X_2) = X_2^2 + \alpha_1 X_2 + (\alpha_1^2 + 3);$$

$$m_{\alpha_3, K_2}(X_3) = X_3 + (-4/3)\alpha_1^2\alpha_2^2 + (-2/3)\alpha_1^2 + (-2/3)\alpha_2^2$$
$$+ (10/3)\alpha_1\alpha_2 + (7/3)\alpha_1 + (7/3)\alpha_2 - 2.$$

Now our procedure for determining the f_i, described before the Structure Theorem, yields

$$f_1(X_1, X_2, X_3) = X_1^3 + 3X_1 + 1;$$

$$f_2(X_1, X_2, X_3) = X_2^2 + X_1 X_2 + (X_1^2 + 3);$$

$$f_3(X_1, X_2, X_3) = X_3 + (-4/3)X_1^2 X_2^2 + (-2/3)X_1^2$$
$$+ (-2/3)X_2^2 + (10/3)X_1 X_2 + (7/3)X_1 + (7/3)X_2 - 2.$$

Hence

$$I = \big(X_1^3 + 3X_1 + 1,$$

$$X_2^2 + X_1 X_2 + (X_1^2 + 3),$$

$$X_3 + (-4/3)X_1^2 X_2^2 + (-2/3)X_1^2 + (-2/3)X_2^2$$

$$+ (10/3)X_1 X_2 + (7/3)X_1 + (7/3)X_2 - 2\big).$$

Now we consider the conditions on $\beta_i = \epsilon(\alpha_i)$ that must hold for ϵ to be an isomorphism $\epsilon\colon K_3 \to K_3$. The only isomorphism from $K_0 = \mathbb{Q}$ to itself is the identity (Exercise 20.6), so we take for $\epsilon_0\colon K \to K$ the identity map $\epsilon_0(k) = k$.

The possible choices for β_1 are the roots of

$$\epsilon_0(m_{\alpha_1, K}(X_1)) = \epsilon_0(X_1^3 + 3X_1 + 1) = X_1^3 + 3X_1 + 1,$$

which are precisely α_1, α_2, and α_3. For each choice α_j, we obtain a different $\epsilon_1\colon K(\alpha_1) \to K(\alpha_j)$, which is the identity on K and sends α_1 to α_j.

The possible choices for β_2, given a choice for β_1, are the roots of

$$\epsilon_1(m_{\alpha_2, K_1}(X_2)) = \epsilon_1(X_2^2 + \alpha_1 X_2 + (\alpha_1^2 + 3)) = X_2^2 + \beta_1 X_2 + (\beta_1^2 + 3).$$

Since this polynomial has degree 2, by Theorem 11.1, there are at most two roots in L, and furthermore we have that $\epsilon_1(m_{\alpha_2, K_1})(X_2)$ divides

$$\epsilon_1(m_{\alpha_2, K})(X_2) = m_{\alpha_2, K}(X_2) = m_{\alpha_1, K}(X_2).$$

(Why?) Using *Maple* or *Mathematica*, we verify that if $\beta_1 = \alpha_1$, then β_2 may take either of the values α_2 and α_3; if $\beta_1 = \alpha_2$, then β_2 may be either α_1 or α_3, and if $\beta_1 = \alpha_3$, then β_2 may equal either α_1 or α_2.

The possible choices for β_3, given choices for β_1 and β_2, are the roots of

$$\epsilon_2(m_{\alpha_3, K_2}(X_3)) = \epsilon_2\big(X_3 + (-4/3)\alpha_1^2\alpha_2^2 + (-2/3)\alpha_1^2 + (-2/3)\alpha_2^2$$
$$+ (10/3)\alpha_1\alpha_2 + (7/3)\alpha_1 + (7/3)\alpha_2 - 2\big)$$
$$= X_3 + (-4/3)\beta_1^2\beta_2^2 + (-2/3)\beta_1^2 + (-2/3)\beta_2^2$$
$$+ (10/3)\beta_1\beta_2 + (7/3)\beta_1 + (7/3)\beta_2 - 2.$$

Because this polynomial is linear, there is only a single root of the polynomial in L. The root is dependent on a choice of β_2 from the possibilities constrained by the choice of β_1. Using *Maple* or *Mathematica*, we verify that if β_1 and β_2 are any two of α_1, α_2, and α_3, the unique choice for β_3 is the remaining α_i.

These possibilities exhaust the possible isomorphisms $\epsilon = \epsilon_3$ from K_3 to K_3. Note that the set $\{\beta_1, \beta_2, \beta_3\}$ is equal to the set $\{\alpha_1, \alpha_2, \alpha_3\}$ in each case. We have determined that there exist precisely six automorphisms of K_3, and each permutes the α_i in a distinct fashion.

Example 17.2. (Compare Example 24.15.) Let α_1, α_2, and α_3 be the three roots of $m_{\alpha_i, \mathbb{Q}} = X^3 - 3X - 1$, where

$$\alpha_1 \approx -1.532, \quad \alpha_2 \approx -0.348, \quad \text{and} \quad \alpha_3 \approx 1.879.$$

Let $K = K_0 = \mathbb{Q}$ and $K_i = \mathbb{Q}(\alpha_1, \ldots, \alpha_i)$ for $i = 1, 2, 3$.

As in the previous example, we calculate the I of the Structure Theorem such that $\mathbb{Q}[X_1, X_2, X_3]/I \cong K_3$. The three minimal polynomials $m_{\alpha_i, K_{i-1}}(X_i)$ are

$$m_{\alpha_1, K_0}(X_1) = X_1^3 - 3X_1 - 1;$$
$$m_{\alpha_2, K_1}(X_2) = X_2 + (\alpha_1^2 - 2);$$
$$m_{\alpha_3, K_2}(X_3) = X_3 + (-\alpha_1^2 + \alpha_1 + 2).$$

Before going on, we can already observe that $[K_3 : \mathbb{Q}] = [K_3 : K_2][K_2 : K_1][K_1 : K] = 3$; hence $K_3 = K_2 = K_1 = \mathbb{Q}(\alpha_1)$. Moreover, since $\deg(\alpha_i) = 3$ for all i, $K_3 = \mathbb{Q}(\alpha_1) = \mathbb{Q}(\alpha_2) = \mathbb{Q}(\alpha_3)$. Putting these facts together, we already know that any isomorphism from K_3 to K_3 is completely determined by the image of α_1. We surmise that we may send α_1 to α_1, α_2, or α_3, but let's check.

The generators of I are

$$f_1(X_1, X_2, X_3) = X_1^3 - 3X_1 - 1;$$

$$f_2(X_1, X_2, X_3) = X_2 + (X_1^2 - 2);$$

$$f_3(X_1, X_2, X_3) = X_3 + (-X_1^2 + X_1 + 2).$$

Hence

$$I = \left(X_1^3 - 3X_1 - 1, \ X_2 + (X_1^2 - 2), \ X_3 + (-X_1^2 + X_1 + 2) \right).$$

Now we consider the conditions on $\beta_i = \epsilon(\alpha_i)$ that must hold for ϵ to be an isomorphism $\epsilon \colon K_3 \to K_3$. The only isomorphism from $K_0 = \mathbb{Q}$ to itself is the identity (Exercise 20.6), so we take for $\epsilon_0 \colon K \to K$ the identity map $\epsilon_0(k) = k$.

The possible choices for β_1 are the roots of

$$\epsilon_0(m_{\alpha_1, K}(X_1)) = \epsilon_0(X_1^3 - 3X_1 - 1) = X_1^3 - 3X_1 - 1,$$

which are precisely α_1, α_2, and α_3. For each choice of α_j, we obtain a different $\epsilon_1 \colon K(\alpha_1) \to K(\alpha_j)$, which is the identity on K and sends α_1 to α_j.

The possible choices for β_2, given a choice for β_1, are the roots of

$$\epsilon_1(m_{\alpha_2, K_1}(X_2)) = \epsilon_1(X_2 + (\alpha_1^2 - 2)) = X_2 + (\beta_1^2 - 2).$$

Since this polynomial is linear, the polynomial has only a single root in L, equal to $-(\beta_1^2 - 2)$. Using *Maple* or *Mathematica*, we verify that if $\beta_1 = \alpha_1$, then $-(\beta_1^2 - 2) = \beta_2$ must be α_2; if $\beta_1 = \alpha_2$, then $\beta_2 = \alpha_3$, and if $\beta_1 = \alpha_3$, then $\beta_2 = \alpha_1$.

The possible choices for β_3, given a choice for β_1 (which then determines the choice for β_2), are similarly constrained to be the roots of

$$\epsilon_2(m_{\alpha_3, K_2}(X_3)) = \epsilon_2(X_3 + (-\alpha_1^2 + \alpha_1 + 2)) = X_3 + (-\beta_1^2 + \beta_1 + 2).$$

Given a choice of β_1, not only β_2 but also β_3 is uniquely determined. What must β_3 be in each case?

These possibilities exhaust the possible isomorphisms ϵ from K_3 to K_3, and the set $\{\beta_1, \beta_2, \beta_3\}$ is equal to the set $\{\alpha_1, \alpha_2, \alpha_3\}$ in each case. We have determined that there exist precisely three automorphisms of K_3, each of which permutes the $\alpha_1, \alpha_2, \alpha_3$ along the cycle $\alpha_1 \to \alpha_2 \to \alpha_3 \to \alpha_1$.

Example 17.3. (Compare Example 24.16.) Let $\alpha_1, \alpha_2, \alpha_3$ be the three roots of $X^3 - 2$, where

$$\alpha_1 \approx 1.260, \quad \alpha_2 \approx -.630 - 1.091i, \quad \text{and} \quad \alpha_3 \approx -.630 + 1.091i.$$

As before, let $K = K_0 = \mathbb{Q}$ and $K_i = \mathbb{Q}(\alpha_1, \ldots, \alpha_i)$ for $i = 1, 2, 3$.

We calculate the minimal polynomials $m_{\alpha_i, K_{i-1}}(X_i)$:

$$m_{\alpha_1, K_0}(X_1) = X_1^3 - 2;$$
$$m_{\alpha_2, K_1}(X_2) = X_2^2 + \alpha_1 X_2 + \alpha_1^2;$$
$$m_{\alpha_3, K_2}(X_3) = X_3 - \alpha_1 - \alpha_2.$$

The final linear polynomial again verifies the fact that once we know where the two roots α_1, α_2 are sent under an automorphism of K_3, we will have determined where the last must be sent.

Since the first polynomial is of degree 3, we know that β_1 may be any of the three roots α_1, α_2, and α_3, and once we choose such a β_1, the fact that the second polynomial is of degree 2 indicates that we will have two choices for β_2. In the end, we have six automorphisms of K_3, and each of these permutes $\alpha_1, \alpha_2, \alpha_3$ in a distinct fashion.

Looking ahead, we will see in section 32 that there is a more enlightening way to understand the six automorphisms of K_3 in this special example. The three roots are in fact $\alpha_1, \omega\alpha_1$, and $\omega^2\alpha_1$, where ω is a primitive third root of unity, with minimal polynomial $X^2 + X + 1$ (over $\mathbb{Q}$ and over $\mathbb{Q}(\alpha_1)$). The six automorphisms correspond to the choices of sending α_1 to $\alpha_1, \omega\alpha_1$, or $\omega^2\alpha_1$ and of sending ω to either ω or ω^2.

18. Isomorphisms from Multiply Generated Fields

Now, building on the Structure Theorem and the examples from the last section, we prove a stable of results on isomorphisms from multiply generated fields to other fields – in particular, isomorphisms that extend an isomorphism of subfields.

18.1. Conditions for Isomorphisms from Multiply Generated Fields

Definition 18.1 (Extension of Isomorphism, Isomorphism over a Field). Let K, L, and M be fields with $K \subset L$ and $K \subset M$, and let $\tau : K \to K$ be an automorphism. An isomorphism

$\epsilon: L \to M$ is said to *extend* τ if ϵ restricts to τ on K, that is, if $\epsilon(k) = \tau(k)$ for every $k \in K$. If ϵ extends the identity automorphism on K, then ϵ is said to be an isomorphism *over K*.

While we want to extend isomorphisms, we prove the following more general result about extending monomorphisms (one-to-one homomorphisms); remember that every nontrivial homomorphism of fields is one-to-one.

Theorem 18.2 (Extension of Monomorphism). *Let K be a subfield of $\mathbb{C}$, L a field extension of K, α an algebraic number, and $\tau: K \to L$ a monomorphism.*

Then τ extends to a monomorphism $\epsilon: K(\alpha) \to L$ if and only if there exists an element $\beta \in L$ that is a root of $\tau(m_{\alpha, K})$.

Proof. First we show that the conditions of the theorem are necessary for the existence of a monomorphism ϵ by assuming that ϵ exists and then deriving the condition. Suppose that ϵ exists, and let $\beta = \epsilon(\alpha)$. Applying ϵ to both sides of the equation $m_{\alpha, K}(\alpha) = 0$, we deduce that $\epsilon(m_{\alpha, K})(\beta) = 0$. (To see this, write $m_{\alpha, K}(X) = \sum c_i X^i$. Then $m_{\alpha, K}(\alpha) = \sum c_i \alpha^i$, and $\epsilon(m_{\alpha, K})(\alpha) = \sum \epsilon(c_i) \alpha^i$. Compare with $\epsilon(m_{\alpha, K})(\beta)$.) Since the coefficients of $m_{\alpha, K}(X)$ lie in K, $\epsilon(m_{\alpha, K}) = \tau(m_{\alpha, K})$. Hence $\beta \in L$ is a root of $\tau(m_{\alpha, K})$ and the condition is satisfied.

Now assume that the condition holds. Let $\tilde{L}$ be the image of K under τ, and consider the onto homomorphism $\varphi: K[X] \to \tilde{L}(\beta) \subset L$ defined by $\varphi(p) = \tau(p)(\beta)$. By hypothesis, $m_{\alpha, K}(X)$ and hence the principal ideal $(m_{\alpha, K})$ lie in the kernel of φ. Suppose that $(m_{\alpha, K}) \neq \ker \varphi$. Because $K[X]$ is a principal ideal domain, there exists $h \in K[X]$ such that $\ker \varphi = (h)$, from which it follows that $h(X)$ divides the irreducible $m_{\alpha, K}(X)$. Now if $\deg h = \deg m_{\alpha, K}$, then h and $m_{\alpha, K}$ differ by a multiplicative constant and $(h(X)) = (m_{\alpha, K}(X))$, a contradiction. Otherwise, h must be a nonzero constant, and hence $(h) = (1)$, so that φ is the zero map, another contradiction. Hence by the First Isomorphism for Rings (3.1), $K[X]/(m_{\alpha, K}(X)) \cong \tilde{L}(\beta)$. Now compare with the isomorphism of Theorem 8.12. $\square$

Note that the extension ϵ of a monomorphism τ as in the last theorem is uniquely determined by τ and β, that is, there is only one $\epsilon: K(\alpha) \to L$ that extends L and sends α to β. Observe also that under the isomorphism in the theorem, an arbitrary expression of the form $\sum c_i \alpha^i$, $c_i \in K$, is sent to $\sum \tau(c_i) \beta^i$.

Theorem 18.3 (Extension of Automorphism to a Multiply Generated Field). *Let K be a subfield of $\mathbb{C}$, $\alpha_1, \alpha_2, \ldots, \alpha_n$ algebraic numbers, L a field extension of K, and $\epsilon_0 = \tau: K \to K$ an automorphism. Let $K_0 = K$ and $K_i = K(\alpha_1, \ldots, \alpha_i)$.*

Then $K(\alpha_1, \ldots, \alpha_n)$ is isomorphic to L under an isomorphism ϵ that extends τ if and only if the Extension of Monomorphism Theorem (18.2) may be applied iteratively, that is, we may iteratively determine for $i = 1, \ldots, n$ homomorphisms $\epsilon_i \colon K_i \to L$ and elements $\beta_i \in L$ such that

- *β_i is a root of $\epsilon_{i-1}(m_{\alpha_i, K_{i-1}})(X)$, for $i = 1, \ldots, n$, and*
- *$L = K(\beta_1, \ldots, \beta_n)$.*

Observe that if the conditions of Theorem 18.3 are satisfied, the $\beta_i \in L$ uniquely determine the isomorphism $\epsilon = \epsilon_n \colon K_n \to L$ with $\epsilon(k) = \tau(k)$ for all $k \in K$ and $\epsilon(\alpha_i) = \beta_i$ for $1 \le i \le n$.

We use Theorem 18.3 to prove the following important corollary: that automorphisms, over the base field, of the splitting field of a polynomial p correspond to permutations of the roots of p. We already observed this phenomenon in section 17, where we also saw that not every permutation of the roots of p will necessarily correspond to an automorphism. We will study this situation closely in Chapter 5. For now, we prove the claim and then prove the Extension of Automorphism to a Multiply Generated Field Theorem.

While we use Theorem 18.3 to prove the corollary, the theorem is not, in fact, necessary, and it is an exercise (20.13) to provide a direct proof. However, we adopt this proof as preparation for the proof of Theorem 18.7 in the next section.

Corollary 18.4 (Automorphisms of Splitting Field Correspond to Permutations). *Let K be a subfield of $\mathbb{C}$ and L the splitting field over K of a polynomial $p \in K[X]$. Then each automorphism ϵ of L extending the identity automorphism $\tau \colon K \to K$ permutes the roots of p in L, and each automorphism is uniquely determined by τ and this permutation.*

Proof. Let $\alpha_1, \ldots, \alpha_n$ be the roots of p. Each is a root of a corresponding monic polynomial $m_{\alpha_i, K}$. Suppose that ϵ is an automorphism of L extending the identity automorphism $\tau \colon K \to K$. By the Extension of Automorphism to a Multiply Generated Field Theorem (18.3), for each i, $\epsilon(\alpha_i) = \beta_i$ for some β_i a root of $\epsilon(m_{\alpha_i, K_{i-1}})$. But $m_{\alpha_i, K_{i-1}}$ divides $m_{\alpha_i, K}$ in $K_{i-1}[X]$ and hence $\epsilon(m_{\alpha_i, K_{i-1}})$ divides $\epsilon(m_{\alpha_i, K}) = \tau(m_{\alpha_i, K}) = m_{\alpha_i, K}$. Then every β_i is a root of $m_{\alpha_i, K}$ and hence a root of p. Now no two roots α_i, α_j may be mapped to the same root of p under ϵ, since ϵ is an automorphism. A one-to-one mapping of a finite set to itself is necessarily onto, and hence a permutation, so we are done. $\square$

Proof of Theorem. First we show that the conditions are necessary. Suppose that such an isomorphism $\epsilon: K(\alpha_1, \ldots, \alpha_n) \to L$ exists. Let $\epsilon_i: K(\alpha_1, \ldots, \alpha_i) \to L$ be the restriction of the homomorphism to K_i. Then each ϵ_i is a monomorphism extending the monomorphism ϵ_{i-1}, and hence by the Extension of Monomorphism Theorem (18.2), there must exist a $\beta_i = \epsilon_i(\alpha_i)$ that is a root of $\epsilon_{i-1}(m_{\alpha_i, K_{i-1}})$. Furthermore, since the image of the isomorphism ϵ is contained in $K(\beta_1, \ldots, \beta_n)$, we must have that $L = K(\beta_1, \ldots, \beta_n)$.

Now we show that the conditions are sufficient. The conditions allow us to apply the Extension of Monomorphism Theorem (18.2) iteratively, yielding monomorphisms $\epsilon_i: K_i \to L$, where each ϵ_i extends ϵ_{i-1} for $i = 1, \ldots, n$. Once we do so, we have a monomorphism $\epsilon = \epsilon_n: K(\alpha_1, \ldots, \alpha_n) \to L$ with image $K(\beta_1, \ldots, \beta_n)$. If the last condition, that $L = K(\beta_1, \ldots, \beta_n)$, holds, then we have an isomorphism. $\qquad\square$

18.2. Isomorphisms of Splitting Fields over Isomorphic Fields

Finally, by taking careful note of the methods we have used so far, we extend these results to the extension of isomorphisms $\tau: K \to K'$ between *different* fields K and K', rather than the extension of automorphisms $\tau: K \to K$. At this point, these results may seem unnecessarily general; however, these powerful results will be fundamental to the construction of the Galois correspondence in Chapter 5. They provide the essential tools to study isomorphisms of splitting fields, in the general context that we will need, and they do not require much additional work.

The main tool is the following:

Theorem 18.5. *Let $\tau: K \to K'$ be an isomorphism of fields, and let $p \in K[X]$ be a polynomial that is irreducible over K. Then $K[X]/(p(X))$ is isomorphic to $K'[X]/(\tau(p)(X))$ under an isomorphism extending τ and sending $X + p(X)$ to $X + \tau(p)(X)$.*

Proof. Let $\epsilon: K[X] \to K'[X]/(\tau(p(X)))$ be the function defined by $\epsilon(f) = \tau(f) + (\tau(p(X)))$. Since τ is a homomorphism from $K[X]$ onto $K'[X]$, this is an onto homomorphism of rings. The kernel $\ker \epsilon$ of ϵ is the set of polynomials $f \in K[X]$ such that $\tau(f) = \tau(p)q'$ for some polynomial $q' \in K'[X]$. Since $p \in \ker \epsilon$, the principal ideal $(p) \subset \ker \epsilon$.

Now we claim that $(p) = \ker \epsilon$, as follows. Because $K[X]$ is a principal ideal domain, there exists $h \in K[X]$ such that $\ker \epsilon = (h)$, from which it follows that $h(X)$ divides $p(X)$. If $h(X)$ is of the same degree as $p(X)$, then they differ by a multiplicative constant and

$(h(X)) = (p(X))$, proving the claim. Otherwise, $h(X)$ is a nonzero constant, so that $(h) = (1)$ and ϵ is the zero map, a contradiction. Hence $\ker \epsilon = (p(X))$.

By the First Isomorphism Theorem for Rings (3.1), $K[X]/(p(X)) \cong K'[X]/(\tau(p)(X))$ under an isomorphism extending τ. □

Now we have two important theorems that follow from the previous result.

Theorem 18.6 (Extension of Isomorphism over Isomorphic Fields). *Let K and K' be subfields of $\mathbb{C}$ isomorphic under an isomorphism $\tau: K \to K'$. Let $p \in K[X]$ be a polynomial that is irreducible over K. Let α be a root of p and α' a root of $\tau(p)$. Then $K(\alpha) \cong K'(\alpha')$ under an isomorphism extending τ and sending α to α'.*

Proof. By the Structure Theorem for Fields Generated by an Algebraic Number (8.12), the fields generated over K and K' by α and α' are isomorphic to the associated quotients $K[X]/(m_{\alpha,K}(X))$ and $K'[X]/(m_{\alpha',K'}(X))$ under isomorphisms that carry α and α' to X, respectively. Applying the previous Theorem 18.5 and transitivity of isomorphism, the theorem is proved. □

Theorem 18.7 (Extension of Isomorphism to Splitting Fields). *Let K and K' be subfields of $\mathbb{C}$ isomorphic under an isomorphism $\tau: K \to K'$. Let $p \in K[X]$ be a polynomial that is irreducible over K. Let L be the splitting field over K of p and L' the splitting field over K' of $\tau(p)$. Then L is isomorphic to L' under an isomorphism extending τ.*

Proof. Let $n = \deg(p)$. Let $\{\alpha_i\}_{i=1}^{n}$ be the n distinct roots of p (Theorem 11.3), and let $K_i = K(\alpha_1, \ldots, \alpha_i)$. We build up an isomorphism by induction on $i \le n$, showing that $K_i \cong K_i'$ under an isomorphism extending τ, where K_i' is generated over K' by i roots of $\tau(p)$, which we will choose as we go. Since $L = K_n$ and L' must therefore be K_n', we will be done.

For the base case, let α_1' be any root of $\tau(p)$. Then $K(\alpha_1) \cong K'(\alpha_1')$ by an isomorphism extending τ, by the Extension of Isomorphism over Isomorphic Fields Theorem (18.6). Now for the inductive step, assume that the statement holds for i: K_i is isomorphic to K_i' under an isomorphism ϵ_i extending τ; we determine K_{i+1}' so that the statement holds for $i + 1$.

Let $\epsilon_i: K_i[X] \to K_i'[X]$ be the isomorphism given by induction, extended to the polynomial rings by acting on the coefficients. By the isomorphism property, since $m_{\alpha_{i+1},K_i}(X)$ is irreducible and divides p, $\epsilon_i(m_{\alpha_{i+1},K_i}(X))$ is irreducible and divides $\epsilon_i(p) = \tau(p)$ in $K_i'[X]$.

Hence all of the roots of $\epsilon_i(m_{\alpha_{i+1}, K_i})$ are roots of $\tau(p)$. Since we are given that L' contains all of the roots of $\tau(p)$, we know that there exists a root, say α'_{i+1}, of $\epsilon_i(m_{\alpha_{i+1}, K_i})$ in L'. Then, by the Extension of Isomorphism over Isomorphic Fields Theorem (18.6), K_{i+1} is isomorphic to $K'_{i+1} = K'_i(\alpha'_{i+1})$. By that theorem, this isomorphism extends the isomorphism ϵ_i, hence also τ, and our proof is finished.   □

Counting the number of isomorphism extensions available in this last theorem will be one of our tasks in sections 24 and 35.

19. Fields and Splitting Fields Generated by Arbitrarily Many Algebraic Numbers

In this section, we step back for a moment before moving on to the Galois correspondence. We explore what the results we have already established tell us about the three field-theoretic properties of being simple, finite, and algebraic.

19.1. Adjoining Arbitrarily Many Algebraic Numbers Leaves a Field Algebraic

Let K be a field consisting only of algebraic numbers. We may certainly consider a field generated over K by an infinite number of distinct algebraic numbers, and often such a field will have infinite degree over K (but not always: see Exercise 20.9). We show here that such a field consists entirely of algebraic numbers, generalizing Theorem 12.12.

In fact, we show more, that if for each adjoined algebraic number, we also adjoin every conjugate of the algebraic number over K, then the resulting field contains each root of the minimal polynomials over K of every one of its elements.

Theorem 19.1. *Let $\{\alpha_i\}_{i \in I}$ be an indexed set of algebraic numbers, $K \subset \mathbb{C}$ a subfield of algebraic numbers, and $L = K(\{\alpha_i\})$. Then each $\beta \in L$ is an algebraic number.*

Moreover, if for each α_i, the complete set of roots of $m_{\alpha_i, K}$ is contained in $\{\alpha_i\}$, then for each $\beta \in L$, the complete set of roots of $m_{\beta, K}$ is contained in L.

Proof. Let β be an element of L. Now β is represented by an arithmetic combination of $\{\alpha_i\}_{i \in I}$ over K. However, by the definition of arithmetic combination, this arithmetic combination contains only a finite number of the α_i, say $\alpha_{i_1}, \ldots, \alpha_{i_k}$, and hence β is an element of $K(\alpha_{i_1}, \ldots, \alpha_{i_k})$. By inductively applying the result that adjoining an algebraic number to a field of algebraic numbers extends the field only by algebraic numbers (Theorem 12.12), β is an algebraic number and we have proved the first statement.

Now each α_{i_j} has only a finite number of conjugates over K, so the set S consisting of each α_{i_j} and its conjugates is finite. Let β' be an arbitrary conjugate of β over K. Then there exists an isomorphism $\epsilon: K(\beta) \cong K(\beta')$ over K by the Extension of Isomorphism over Isomorphic Fields Theorem (18.6).

Now consider $K(\beta)(S)$ and $K(\beta')(S)$: these are the splitting fields of the product of minimal polynomials $\prod_{j=1}^{k} m_{\alpha_{i_j},K}(X)$ over $K(\beta)$ and $K(\beta')$, respectively. By a generalization of the Extension of Isomorphism to Splitting Fields Theorem (18.7) in Exercise 20.14, ϵ extends to an isomorphism $\bar{\epsilon}: K(\beta)(S) \to K(\beta')(S)$ between these splitting fields.

Now since β is represented by an arithmetic combination of $\{\alpha_{i_1}, \ldots, \alpha_{i_k}\} \subset S$,

$$K(\beta)(S) = K(S)(\beta) = K(S).$$

Suppose β' is not an element of $K(S)$. Then $m_{\beta',K(S)}$ is a polynomial of degree at least 2, implying that $[K(\beta')(S) : K(S)] > 1$ by Theorem 8.6. Moreover,

$$[K(\beta')(S) : K] = [K(\beta')(S) : K(S)][K(S) : K] > [K(S) : K]$$

by Exercise 13.8. However, since dimension is an invariant under isomorphism of vector spaces, the existence of $\bar{\epsilon}$ implies that $[K(\beta')(S) : K] = [K(S) : K]$. Hence we have reached a contradiction, and $\beta' \in K(S) \subset L$. $\square$

This theorem holds in more generality, given the definition of *algebraic* in the next section, and it is an exercise (20.18) to adapt the proof.

19.2. Properties of Characteristic Zero Fields: Simple versus Finite versus Algebraic Extensions

Now we introduce the following definitions from the general theory of fields that encapsulate important properties of field extensions.

Definition 19.2 (Generated Field). A field extension L/K is *generated by a set $S \subset L$ over K* if the smallest subfield of L containing K and S is L. If so, we write $L = K(S)$. We say that the field L is *finitely generated over K* if $L = K(S)$ for some finite set S.

The preceding definition generalizes our previous definitions of fields generated by one (Definition 8.2) or more (Definition 16.1) algebraic numbers over a subfield of $\mathbb{C}$.

Definition 19.3 (Simple Extension). A field extension L/K is *simple* if $L = K(\alpha)$ for some $\alpha \in L$. We say that α is a *primitive element* of the extension L/K.

Note that α is only an element of L, not necessarily a number. An example of a simple extension that is generated by a nonnumeric object is the following:

Example 19.4 (Transcendental Simple Extension). Let $\mathbb{Q}(t)$ denote the *field of rational functions in one variable* with coefficients in $\mathbb{Q}$:

$$\mathbb{Q}(t) = \left\{ \frac{p(t)}{q(t)} \;\middle|\; p, q \in \mathbb{Q}[t], \; q(t) \neq 0 \right\},$$

where, as usual, two elements p/q and p'/q' are declared equivalent if $pq' = p'q$, and with addition, multiplication, and equivalence of fractions defined in the usual ways. Then, as a field extension, $\mathbb{Q}(t)$ is generated over $\mathbb{Q}$ by t.

Definition 19.5 (Finite Extension). A field extension L/K is *finite* if $[L : K] < \infty$.

Definition 19.6 (Algebraic Element, Extension). Let L/K be a field extension. An element $\beta \in L$ is *algebraic over K* if β is the root of a polynomial in $K[X]$. We say that L/K is *algebraic* if each $\beta \in L$ is algebraic over K.

Note that this last definition offers an alternative definition for an algebraic number: an algebraic number is a complex number that is algebraic over $\mathbb{Q}$.

In this chapter, we have concerned ourselves only with subfields of $\mathbb{C}$, and we close the theoretical portion of this chapter by restating, in terms of the properties above, some of the results we have proven about general subfields of $\mathbb{C}$.

Theorem 19.7 (Finite = Simple and Algebraic). *Let L be a subfield of $\mathbb{C}$. Then the field extension $L/\mathbb{Q}$ is finite if and only if it is simple and algebraic.*

It is an exercise (20.20) to prove this theorem in somewhat more generality.

Proof. If $L/\mathbb{Q}$ is both simple and algebraic, then L is generated over $\mathbb{Q}$ by a single algebraic number α, and we saw long ago that $\mathbb{Q}(\alpha)/\mathbb{Q}$ is finite (Theorem 8.6).

For the other direction, suppose that $L/\mathbb{Q}$ is finite. If there exists an element $l \in L$ that is not an algebraic number, then the infinite set $\{l^i\}_{i=0}^{\infty}$ will be linearly independent over $\mathbb{Q}$, forcing $[\mathbb{Q}(l) : \mathbb{Q}]$ to be infinite. If $[\mathbb{Q}(l) : \mathbb{Q}]$ is infinite, however, then we have a contradiction by the Tower Law (Exercise 13.8), since $[L : \mathbb{Q}] = [L : \mathbb{Q}(l)][\mathbb{Q}(l) : \mathbb{Q}]$. Hence $L/\mathbb{Q}$ is algebraic.

Now we prove that if $L/\mathbb{Q}$ is finite, then it is simple. Let $\{x_i\}$ be a basis for the vector space L over $\mathbb{Q}$. Then $L = \mathbb{Q}(x_1, \ldots, x_n)$, with each x_i algebraic by the preceding paragraph, and after an inductive application of the result that generation by two algebraic numbers is generation by one (Theorem 16.2), we deduce that $L = \mathbb{Q}(y)$ for some $y \in L$.   □

We have established that for an extension $L/\mathbb{Q}$, the property of being finite is equivalent to the conjunction of the properties of simple and algebraic. Now we go on to show with examples that there are no other relations.

The complex number π is known to be the root of no polynomial with rational coefficients; in other words, it is *transcendental* ([54]; see also [21, §III.14] or [25, Chap. 24]). Hence $\mathbb{Q}(\pi)$ is a field extension of $\mathbb{Q}$ that is simple but not algebraic.

Similarly, since $\{\pi^i\}_{i=0}^{\infty}$ forms an infinite set of linearly independent elements of $\mathbb{Q}(\pi)$ over $\mathbb{Q}$, we have that $\mathbb{Q}(\pi)/\mathbb{Q}$ is a field extension that is simple but not finite.

Now by Exercise 20.16, the field $L = \mathbb{Q}(\{\sqrt{p} : p \in \mathbb{P}\})$, where $\mathbb{P}$ is the set of prime numbers in $\mathbb{Z}$, is an extension of $\mathbb{Q}$ of infinite dimension over $\mathbb{Q}$, hence not finite. Since each $\sqrt{p}$ is algebraic, and generation over a field of algebraic numbers by an arbitrary number of algebraic numbers results in an algebraic field, the entire field L is algebraic (Theorem 19.1). Now a field extension generated by a single algebraic number is finite; hence the algebraic field extension $L/\mathbb{Q}$ cannot be simple. Therefore $L/\mathbb{Q}$ is algebraic but neither simple nor finite.

We now define a number field.

Definition 19.8 (Number Field). A *number field* is a finite extension of $\mathbb{Q}$.

Number fields are the fields of primary interest in this text and, by the previous theorem (19.7), these include all of the fields we have studied that are generated by adjoining a finite number of algebraic numbers to $\mathbb{Q}$.

20. Exercise Set 1

A.

20.1* Let D be an integral domain. Give a definition for the polynomial ring $D[X_1, \dots, X_n]$ in n variables over D, and prove that these polynomial rings are integral domains.

20.2 Prove that an $s \in \mathbb{Q}$ exists satisfying the requirements in the proof of Theorem 16.2.

20.3 Let $M = \mathbb{Q}(\sqrt{2}, \sqrt{3})$. Study M as follows: (a) Prove that $[M : \mathbb{Q}] = 4$; (b) find a basis S for M over $\mathbb{Q}$ and prove that S is a basis; (c) find an element γ such that $M = \mathbb{Q}(\gamma)$; (d) express $\sqrt{2}$ and $\sqrt{3}$ in reduced form in γ over $\mathbb{Q}$.

20.4* Let S be a finite set of algebraic numbers. Let $\mathbb{Q}(S)$ denote the field $\mathbb{Q}$ with each number in S adjoined. Prove that $\mathbb{Q}(S)$ is a simple extension of $\mathbb{Q}$.

20.5* Let K be a subfield of $\mathbb{C}$ and $\alpha_1, \alpha_2, \ldots, \alpha_n$ algebraic numbers. Prove that every element of $K(\alpha_1, \ldots, \alpha_n)$ may be expressed as a sum of K-multiples of products of powers of the α_i, where the power of α_i does not exceed $\deg_{K(\alpha_1, \ldots, \alpha_{i-1})}(\alpha_i) - 1$. (For a constructive approach, first show that $K(\alpha_1, \ldots, \alpha_n) = K[\alpha_1, \ldots, \alpha_n]$, by induction on n. Then work again by induction, dividing an arbitrary element of the ring by $m_{\alpha_n, K(\alpha_1, \ldots, \alpha_{n-1})}$.) Find a basis for $K(\alpha_1, \ldots, \alpha_n)$ over K consisting only of products of powers of the α_i, and give an algorithm for producing a *reduced form in* $\alpha_1, \ldots, \alpha_n$ for elements of L.

20.6* Show that if $K = \mathbb{Q}$, then every isomorphism between field extensions of K is an isomorphism over K, that is, is the identity on K. (Hint: Where does 1 go? Consider quotients carefully, with details.) Generalize, with proof, the result to any K for which the smallest subfield of K containing the multiplicative identity element (called the *prime subfield*) is K itself.

20.7 Let α be a root of $X^n - a$, where $a \in \mathbb{C}$. Prove that $X^n - a$ factors over $\mathbb{C}$ as $\prod_{j=0}^{n-1} (X - \omega^j \alpha)$, where $\omega = e^{2\pi i/n}$, a primitive nth root of 1. Now suppose $\alpha \in K(\omega)$ for K a subfield of $\mathbb{C}$. Prove that the factorization occurs over $K(\omega)$.

20.8* Prove Theorem 16.2 with the more general hypothesis that α, β, and γ are elements of $\mathbb{C}$ that are algebraic over K in the sense of Definition 19.6.

20.9* For each $k \in \mathbb{Q}$, let α_k denote one of the two roots of the polynomial $X^2 - 2kX + (k^2 - 2)$. Determine the dimension of $\mathbb{Q}(\{\alpha_k : k \in \mathbb{Q}\})$ over $\mathbb{Q}$.

B.

20.10 Prove that given a homomorphism $\varphi : K \to L$ of arbitrary fields and a finite set $\{\beta_i\}_{i=1}^n$ of elements of L, there exists a unique homomorphism $\tilde{\varphi} : K[X_1, X_2, \ldots, X_n] \to L$ extending φ and satisfying $\tilde{\varphi}(X_i) = \beta_i$.

20.11 Let K be a subfield of $\mathbb{C}$. Let $p \in K[X]$ be a polynomial that factors over K as $p_1 p_2 \cdots p_k$, and L the splitting field of p over K. Find and prove lower and upper bounds for $[L : K]$ in terms of the set $\{\deg(p_i)\}$. Make your bounds as sharp as you can given this generality.

20.12 ([13, Prop. 4.1]) Let K be a subfield of $\mathbb{C}$, $p \in K[X]$ a polynomial, and L/K the splitting field over K of p. Fill in the details to complete this proof sketch that $[L : K]$ divides $n!$, where $n = \deg p$: Proceed by induction on the degree of the polynomial. First assume that p is irreducible over K. Let α be an arbitrary root of p and consider the subfield $L' = K(\alpha)$. Then since L is the splitting field of $q(X) = p(X)/(X - \alpha)$ over L', $[L : L']$ divides $(\deg q)! = (n-1)!$. Since $\deg p = n$, $[L : K]$ divides $(n-1)! \cdot n = n!$. On the other hand, if p is reducible with an irreducible factor q of degree n' and L' is the splitting field of q over K, then $[L' : K]$ divides $(n')!$. Moreover L is a splitting field of p/q over L', hence $[L : L']$ divides $(n - n')!$. But then $[L : K]$ divides $(n')!(n - n')!$, which divides $n!$.

20.13* Find a direct proof of Corollary 18.4.

20.14* Generalize the Extension of Isomorphism to Splitting Fields Theorem (18.7) to splitting fields of an arbitrary nonconstant polynomial $p \in K[X]$. Use the same proof, carefully.

20.15 Prove that $\mathbb{Q}(t) \cong \mathbb{Q}(\pi)$ as fields.

20.16* Find a basis B for $L = \mathbb{Q}(\{\sqrt{p} \mid p \in \mathbb{P}\})$ over $\mathbb{Q}$, where $\mathbb{P}$ is the set of prime numbers in $\mathbb{Z}$, and prove that B is a basis for L over $\mathbb{Q}$. Then, by Theorem 19.1, we know that L is algebraic, giving us an infinite algebraic extension of $\mathbb{Q}$. (Hint: Index the primes in $\mathbb{P}$ and then prove that $\sqrt{p_i}$ is not in $\mathbb{Q}(\sqrt{p_1}, \ldots, \sqrt{p_{i-1}})$ for each $2 \le i \le n$ by induction, first showing that if $(x + y\sqrt{p_{i-1}})^2 = p_i$ where $x, y \in \mathbb{Q}(\sqrt{p_1}, \ldots, \sqrt{p_{i-2}})$ then $xy = 0$. Once the dimension of $\mathbb{Q}(\sqrt{p_1}, \ldots, \sqrt{p_i})$ is found, consider how a basis is related to a reduced form.) For a note on this problem, see [36].

20.17 Show that finite and algebraic are transitive properties of field extensions. In other words, if L/K is a finite extension and M/L is a finite extension, then M/K is a finite extension, and if L/K is an algebraic extension and M/L is an algebraic extension, then M/K is an algebraic extension. (Hint: For the latter, start with the fact that if $\alpha \in M$ is algebraic over L, it is in fact algebraic over a finite extension of K, namely the one generated over K by the coefficients of $m_{\alpha, L}$.)

20.18* Prove Theorem 19.1, replacing the notion of being an algebraic number with that of being algebraic over K.

20.19 Generalize (with proof) the Structure Theorem for Fields Generated by an Algebraic Number (8.12) for K an arbitrary field and $L = K(\alpha)$ an algebraic extension. You will need to do the same for Proposition 7.2, Theorem 8.6, and Exercise 10.17 along the way.

20.20* Generalize (with proof) the Finite = Simple and Algebraic Theorem (19.7), replacing $\mathbb{Q}$ by an arbitrary subfield K of $\mathbb{C}$ and the notion of an algebraic number with that of an element algebraic over K. (You will need first to formulate and then to prove analogues of the following: Theorem 8.6; the fact that generation over a field of algebraic numbers by an algebraic number results in an algebraic field, Theorem 12.12; and the fact that generation by two algebraic numbers is generation by one, Theorem 16.2.)

20.21 Our work in this chapter has focused on the case when the field extension has $\mathbb{Q}$ for a base field. Examine the following field extensions and prove or disprove the properties of simple, algebraic, and finite for each of them.

(1) $\mathbb{Q}(\sqrt{\pi})/\mathbb{Q}(\pi)$;

(2) $\mathbb{Q}(\sqrt{\pi + 2})/\mathbb{Q}(\pi^2)$;

(3) the splitting field of $X^5 - 1$ over $\mathbb{Q}(\pi)$;

(4) $\mathbb{Q}(\pi, \{\sqrt[n]{\pi} \mid n \geq 2\})/\mathbb{Q}(\pi)$;

(5) $\mathbb{Q}(\pi, \{\sqrt[n]{2} \mid n \geq 2\})/\mathbb{Q}(\pi)$;

(6) $\mathbb{Q}(t, \pi)/\mathbb{Q}(t)$.

21. Computation in Multiply Generated Fields: *Maple* and *Mathematica*

In this section, we examine how the package `AlgFields` implements functions for working with fields generated by several algebraic numbers over the rational numbers.

21.1. Declaring a Field

In order to work with a field generated by several algebraic numbers α_i over the field of rational numbers, we first declare the field to *Maple* or *Mathematica*. We provide a name for each of the generating algebraic numbers α_i. We use these names as the variables for the defining polynomials p_i, which are polynomials associated to the algebraic numbers α_i.

Normally these polynomials are minimal polynomials over $K_{i-1} = \mathbb{Q}(\alpha_1, \ldots, \alpha_{i-1})$. Now if the degree of the polynomial is greater than one, the polynomial does not uniquely determine the algebraic number, and we may, if we choose, specify a number that refers uniquely to the particular root. As before, this root number corresponds, in *Maple*, to the number following `index=` in the `RootOf` function, and, in *Mathematica*, to the second argument of the `Root` function, where in both cases, the first argument is the given polynomial p_i. If the root number is not specified, it is taken to be 1.

The package `AlgFields` permits the polynomial p_i associated to α_i to be reducible over K_{i-1}, in order to accommodate the case when the user wishes to specify an algebraic number by its minimal polynomial over $\mathbb{Q}$, instead of K, and a particular root number.

The precise syntax is as follows. Using the function `FDeclareField`, we pass to the package the name of the field and then (ordered within square brackets in *Maple* or within braces in *Mathematica*), the p_i of the algebraic numbers α_i, using the name for α_i for the indeterminate in p_i. We optionally specify root numbers by passing an ordered list (again in brackets or braces) of these numbers as the second argument.

For instance, we declare a field K1 to be $\mathbb{Q}(\alpha, \beta)$, where $m_{\alpha,\mathbb{Q}} = X^4 - 10X^2 + 20$ and $m_{\beta,\mathbb{Q}} = \beta^2 + \beta + \alpha$, choosing particular root numbers, as follows:

```
> FDeclareField(K1,[
>      alpha^4-10*alpha^2+20,
>      beta^2+beta+alpha],[4,2]);

----Details of field K1 ----

  Algebraic Numbers: [alpha, beta]

  Dimension over Q: 8

  Minimal Polynomials: [

    alpha^4-10*alpha^2+20,

    beta^2+beta+alpha]

  Root Approximations: [

    -2.6899940478558293078,

    -2.2146410842668588372]

----
```

```
In[1]:= FDeclareField[K1, {α^4 - 10α^2 + 20,

            β^2 + β + α}, {1, 1}]

----Details of field K1 ----

  Algebraic Numbers: {α, β}

  Dimension over Q: 8

  Minimal Polynomials: {

    20 - 10 * α^2 + α^4,

    α + β + β^2}

  Root Approximations: {

    -2.6899940478558293078,

    -2.2146410842668588372}

----
```

Note that the field has been declared with respect to particular roots: the root $\alpha \approx -2.690$ of $X^4 - 10X^2 + 20$ and the root $\beta \approx -2.215$ of $X^2 + X + \alpha$.

We declare another field K2 to be $\mathbb{Q}(\gamma, \delta)$, where γ and δ are two particular roots of $X^3 - 2$. Here we use a polynomial for δ that is reducible over $\mathbb{Q}(\gamma)$.

```
>  FDeclareField(K2,[gam^3-2,del^3-2],
>      [1,3]);

----Details of field K2 ----

  Algebraic Numbers: [gam, del]

  Dimension over Q: 6

  Minimal Polynomials: [gam^3-2,
    del^2+del*gam+gam^2]

  Root Approximations: [

    1.2599210498948731648,

    -.62996052494743658240-

    1.0911236359717214036*I]

----
```

```
In[2]:= FDeclareField[K2, {γ^3 − 2, δ^3 − 2}, {1, 2}]

----Details of field K2 ----

  Algebraic Numbers: {γ, δ}

  Dimension over Q: 6

  Minimal Polynomials: {-2 + γ^3,

    γ^2 + γ * δ + δ^2}

  Root Approximations: {

    1.2599210498948731648,

    -0.62996052494743658238-

    1.0911236359717214036 I}

----
```

Now we may additionally want to work with fields generated by one or more algebraic numbers over a previously declared field. We do so with the function FDeclareExtensionField, the arguments of which take the same format as FDeclareField, except that we insert the name of the previously declared field as the second argument.

We declare a field L1 to be K1(ϵ), where $m_{\epsilon, \mathbb{Q}} = X^4 - 10X^2 + 20$ and $\epsilon \approx -1.66$.

```
>  FDeclareExtensionField(L1,K1,
>      [eps^4-10*eps^2+20],[3]);

----Details of field L1 ----
  Algebraic Numbers: [alpha, beta, eps]
  Dimension over Q: 8
  Minimal Polynomials: [
    alpha^4-10*alpha^2+20,
    beta^2+beta+alpha,
    eps+3*alpha-1/2*alpha^3]
  Root Approximations: [
    -2.6899940478558293078,
    -2.2146410842668588372,
    -1.6625077511098137141]

----
```

```
In[3]:= FDeclareExtensionField[L1, K1,
    {ε^4 − 10ε^2 + 20}, {2}]

----Details of field L1 ----
  Algebraic Numbers: {α, β, ε}
  Dimension over Q: 8
  Minimal Polynomials: {
    20 - 10 * α^2 + α^4,
    α + β + β^2, 3 * α - α^3/2 + ε}
  Root Approximations: {
    -2.6899940478558293078,
    -2.2146410842668588372,
    -1.6625077511098137144}

----
```

21.2. Declaring a Splitting Field

To work with a splitting field over $\mathbb{Q}$, we use the function `FDeclareSplittingField` to set up the field in `AlgFields`. Similarly, to work within a splitting field over another declared field, we use the function `FDeclareSplittingExtensionField`. In both cases, we pass to the function an irreducible polynomial over the base field, and the function adjoins each of the roots, in order as determined by *Maple* or *Mathematica*, to the base field to declare the splitting field. A list that names the roots in order may optionally be passed to the function; otherwise, unique names of the form rn are chosen.

We declare a field `KS` as the splitting field of $X^5 - 2$ with default roots r1, r2, r3, r4, and r5.

```
>  FDeclareSplittingField(KS,X^5-2);

----Details of field KS ----

  Algebraic Numbers: [r1, r2, r3, r4, r5]

  Dimension over Q: 20

  Minimal Polynomials: [r1^5-2,
     r2^4+r1*r2^3+r1^2*r2^2+
        r1^3*r2+r1^4,
     -1/2*r1^4*r2^2+r3,
     -1/2*r2^3*r1^3+r4,
     r5+r2+r1+1/2*r2^3*r1^3+
        1/2*r1^4*r2^2]

  Root Approximations: [

     1.1486983549970350068,

     .35496731310463012599+

        1.0924770557774537267*I,

     -.92931649060314762945+

        .67518795239988108310*I,

     -.92931649060314762950-

        .67518795239988108320*I,

     .35496731310463012615-

        1.0924770557774537266*I]

----
```

```
In[4]:= FDeclareSplittingField[KS, X^5 - 2]

----Details of field KS ----

  Algebraic Numbers: {r1, r2, r3, r4, r5}

  Dimension over Q: 20

  Minimal Polynomials: {-2 + r1^5,

     r1^4 + r1^3 * r2 + r1^2 * r2^2 +

        r1 * r2^3 + r2^4,

     r1 + r2 + (r1^4 * r2^2)/2 +

        (r1^3 * r2^3)/2 + r3,

     -(r1^3 * r2^3)/2 + r4,

     -(r1^4 * r2^2)/2 + r5}

  Root Approximations: {

     1.1486983549970350068,

     -0.92931649060314762939 -

        0.67518795239988108308 I,

     -0.92931649060314762939 +

        0.67518795239988108308 I,

     0.35496731310463012599 -

        1.0924770557774537267 I,

     0.35496731310463012599 +

        1.0924770557774537267 I}

----
```

We declare an extension field of K2 as the splitting field over K2 of $X^3 - 3$.

```
>   FDeclareSplittingExtensionField(K2S,
>      K2, X^3-3,[th1,th2,th3]);
----Details of field K2S ----
   Algebraic Numbers: [gam, del, th1,
      th2, th3]
   Dimension over Q: 18
   Minimal Polynomials: [gam^3-2,
      del^2+del*gam+gam^2,
      th1^3-3,
      th2+(1+1/2*del*gam^2)*th1,
      -1/2*th1*del*gam^2+th3]
   Root Approximations: [
      1.2599210498948731648,
      -.62996052494743658240-
         1.0911236359717214036*I,
      1.4422495703074083823,
      -.72112478515370419110+
         1.2490247664834064795*I,
      -.72112478515370419120-
         1.2490247664834064795*I]
----
```

```
In[5]:= FDeclareSplittingExtensionField[
           K2S, K2, X^3 - 3, {θ1,θ2,θ3}]
----Details of field K2S ----
   Algebraic Numbers: {γ, δ, θ1, θ2, θ3}
   Dimension over Q: 18
   Minimal Polynomials: {-2 + γ^3,
      γ^2 + γ * δ + δ^2,
      -3 + θ1^3,
      -(γ^2 * δ * θ1)/2 + θ2,
      (1 + (γ^2 * δ)/2) * θ1 + θ3}
   Root Approximations: {
      1.2599210498948731648,
      -0.62996052494743658238 -
         1.0911236359717214036 I,
      1.4422495703074083823,
      -0.72112478515370419116 -
         1.2490247664834064794 I,
      -0.72112478515370419116 +
         1.2490247664834064794 I}
----
```

21.3. Reduced Forms

Just as with fields declared with a single algebraic number, we may use the function FSimplifyE to bring an arithmetic combination without quotients into reduced form over a multiply generated field, and when a quotient is required, we multiply by the inverse of the denominator, using the function FInvert to compute the inverse.

We reduce the expression $\alpha^{10} + (\alpha + \beta + \epsilon)^7$ in the field L1.

```
>   FSimplifyE(alpha^10+
>      (alpha+beta+eps)^7,L1);
```

$$(17744\,\alpha^2 + 16809\,\alpha - \frac{12189}{2}\,\alpha^3 - 49069)\,\beta +$$
$$51350 + 7160\,\alpha^3 - 19819\,\alpha - 18515\,\alpha^2$$

```
In[6]:= FSimplifyE[α^10+
           (α + β + ε)^7, L1]
```

$$Out[6]= 51350 - 19819\,\alpha - 18515\,\alpha^2 + 7160\,\alpha^3 +$$
$$\left(-49069 + 16809\,\alpha + 17744\,\alpha^2 - \frac{12189\,\alpha^3}{2}\right)\beta$$

We compute the reduced form of the inverse of $\beta\alpha$ in L1.

> ` FInvert(beta*alpha,L1);`

$$\left(\frac{\alpha^2}{20} - \frac{1}{2}\right)\beta + \frac{\alpha^2}{20} - \frac{1}{2}$$

$In[7]:=$ FInvert$[\beta * \alpha, L1]$

$Out[7]=$ $-\frac{1}{2} + \frac{\alpha^2}{20} + \left(-\frac{1}{2} + \frac{\alpha^2}{20}\right)\beta$

Now we put the expression $(\beta\alpha^2)/(\beta\alpha)$ in reduced form.

> ` FSimplifyE((beta*alpha^2)*`
> `   FInvert(beta*alpha,L1),L1);`

$$\alpha$$

$In[8]:=$ FSimplifyE$[(\beta * \alpha^2) * $ FInvert$[\beta * \alpha, L1], L1]$

$Out[8]=$ α

Just as before, we may use the function `FSimplifyP` to reduce the coefficients (without quotients) of a polynomial in one variable over the declared field.

> ` FSimplifyP(X^2+eps^15*X-beta/5,X,L1);`

$$X^2 + (376000\,\alpha^3 - 2720000\,\alpha)\,X - \frac{\beta}{5}$$

$In[9]:=$ FSimplifyP$[X^2 + \epsilon^{15}X - \beta/5, X, L1]$

$Out[9]=$ $X^2 + X(-2720000\,\alpha + 376000\,\alpha^3) - \frac{\beta}{5}$

21.4. Factoring Polynomials over a Field

The function `FFactor` factors a polynomial in one variable over fields declared with multiple algebraic numbers, so long as the polynomial is expressed with coefficients without quotients.

> ` FFactor(X^5-beta^2*X^4-10*X^3+`
> `   10*beta^2*X^2+20*X-20*beta^2,L1);`

$$(X + \beta + \alpha)(X + \alpha)(X - \alpha)$$
$$\left(X - 3\alpha + \frac{1}{2}\alpha^3\right)\left(X + 3\alpha - \frac{1}{2}\alpha^3\right)$$

$In[10]:=$ FFactor$[X^5 - \beta^2X^4 -$
$10X^3 + 10\beta^2X^2 + 20X - 20\beta^2, L1]$

$Out[10]=$ $\frac{1}{4}(X - \alpha)(X + \alpha)$

$(2X + 6\alpha - \alpha^3)(2X - 6\alpha + \alpha^3)(X + \alpha + \beta)$

21.5. The Division Algorithm and Reduced Forms

As before, polynomial quotients and remainders may be found using `FPolynomial` `Quotient` and `FPolynomialRemainder`. The Division Algorithm over fields declared with multiple algebraic numbers is used in the implementation of `FSimplifyE` and `FSimplifyP`, which we saw above.

```
> FPolynomialQuotient(
>    X^3+alpha*X+beta,X-eps,X,L1);
```

$$X^2 + \left(-3\alpha + \frac{1}{2}\alpha^3\right)X + \alpha - \alpha^2 + 10$$

In[11]:= FPolynomialQuotient[$X^3 + \alpha * X + \beta$,
$X - \epsilon, X, L1$]

Out[11]= $10 + X^2 + \alpha - \alpha^2 + X\left(-3\alpha + \dfrac{\alpha^3}{2}\right)$

```
> FPolynomialRemainder(
>    X^3+alpha*X+beta,X-eps,X,L1);
```

$$\beta + 2\alpha^2 + 3\alpha^3 - 20\alpha - 10$$

In[12]:= FPolynomialRemainder[$X^3 + \alpha * X + \beta$,
$X - \epsilon, X, L1$]

Out[12]= $-10 - 20\alpha + 2\alpha^2 + 3\alpha^3 + \beta$

21.6. The Euclidean Algorithm and Inverses

The functions for the computation of polynomial greatest common divisors, `FPolynomialGCD` and `FPolynomialExtendedGCD`, work for these fields as well; indeed, the Euclidean Algorithm is used in the implementation of `FInvert`, which we saw above.

```
> FPolynomialGCD(
>    2*X^5+X^4*(2+2*beta-2*eps)+
>    X^3*(2*beta-2*eps)-
>    2*X^2*beta+alpha*(-2*beta+2*eps)+
>    X*(-2*alpha-2*beta^2+2*beta*eps),
>    2*X^6+X^5*(2+2*alpha)+
>    X^4*(2*alpha+2*beta-2*eps)-
>    2*X^3*eps+X^2*alpha*(-2-2*beta)+
>    X*(-2*alpha^2-2*beta^2+2*beta*eps)+
>    alpha*(-2*beta+2*eps),X,L1);
```

$$X^4 + X^3 - X\beta - \alpha$$

In[13]:= FPolynomialGCD[
$2X^5 + X^4(2 + 2\beta - 2\epsilon)+$
$X^3(2\beta - 2\epsilon)$
$-2X^2\beta + \alpha(-2\beta + 2\epsilon)+$
$X(-2\alpha - 2\beta^2 + 2\beta\epsilon),$
$2X^6 + X^5(2 + 2\alpha)+$
$X^4(2\alpha + 2\beta - 2\epsilon)-$
$2X^3\epsilon + X^2\alpha(-2 - 2\beta)+$
$X(-2\alpha^2 - 2\beta^2 + 2\beta\epsilon)+$
$\alpha(-2\beta + 2\epsilon), X, L1]$

Out[13]= $X^3 + X^4 - \alpha - X\beta$

21.7. Representing Algebraic Numbers and Finding Minimal Polynomials and Factors

Just as with fields generated by a single algebraic number, we may move between two ways of specifying an element δ of a field generated by several algebraic numbers over $\mathbb{Q}$: either as an arithmetic combination of the generating algebraic numbers of the field, or as a particular root of a polynomial.

Here we find the minimal polynomial and particular root number of $\alpha + \beta - \epsilon$ in the field L1, using FMinPoly and FRootNumber.

```
> FMinPoly(alpha+beta-eps,X,L1);
```

$$-580 + 3160\,X - 410\,X^2 + 380\,X^3 +$$

$$391\,X^4 - 76\,X^5 - 34\,X^6 + 4\,X^7 + X^8$$

In[14]:= FMinPoly[$\alpha + \beta - \epsilon$, X, L1]

Out[14]= $-580 + 3160\,X - 410\,X^2 +$

$$380\,X^3 + 391\,X^4 - 76\,X^5 - 34\,X^6 + 4\,X^7 + X^8$$

```
> FRootNumber(%,alpha+beta-eps,L1);
```

$$4$$

In[15]:= FRootNumber[%, $\alpha + \beta - \epsilon$, L1]

Out[15]= 3

Going in the other direction, we find an arithmetic combination in L1 of $\delta = \alpha + \beta - \epsilon$ using FFindFactorRt. (As we observed, in *Maple* the algebraic number δ is the "fourth" root of its minimal polynomial, while in *Mathematica* it is the "third.")

```
> FFindFactorRt(
>     -580+3160*X-410*X^2+380*X^3+391*X^4-
>     76*X^5-34*X^6+4*X^7+X^8,
>     RootOf(-580+3160*X-410*X^2+380*X^3+
>     391*X^4-76*X^5-34*X^6+4*X^7+X^8,
>     index=4),L1);
```

$$X - \beta - 4\,\alpha + \frac{1}{2}\,\alpha^3$$

In[16]:= FFindFactorRt[

$$X^8 + 4X^7 - 34X^6 - 76X^5 + 391X^4 +$$

$$380X^3 - 410X^2 + 3160X - 580,$$

Root[$X^8 + 4X^7 - 34X^6 - 76X^5 + 391X^4$

$$+380X^3 - 410X^2 + 3160X - 580,$$

3], L1]

Out[16]= $2\,X - 8\,\alpha + \alpha^3 - 2\,\beta$

```
> FSimplifyP(
>     (X-beta-4*alpha+1/2*alpha^3)-
>     (X-(alpha+beta-eps)),X,L1);
```

$$0$$

$$In[17]:= \text{FSimplifyP}[(2\,X - 8\,\alpha + \alpha^3 - 2\,\beta)-$$
$$2(X - (\alpha + \beta - \epsilon)), \text{X, L1}]$$

Out[17]= 0

Next we determine that another particular root (the "fifth" in *Maple*, the "first" in *Mathematica*) does not lie in L1, by finding that the corresponding factor of the minimal polynomial has degree greater than one.

```
> FFindFactorRt(
>     -580+3160*X-410*X^2+380*X^3+391*X^4-
>     76*X^5-34*X^6+4*X^7+X^8,
>     RootOf(-580+3160*X-410*X^2+
>     380*X^3+391*X^4-76*X^5-
>     34*X^6+4*X^7+X^8,index=5),
>     L1);
```

$$X^2 + X + 4\,X\alpha - X\alpha^3 - 10 - \alpha + 4\alpha^2$$

$$In[18]:= \text{FFindFactorRt}[$$
$$X^8 + 4X^7 - 34X^6 - 76X^5 + 391X^4+$$
$$380X^3 - 410X^2 + 3160X - 580,$$
$$\text{Root}[X^8 + 4X^7 - 34X^6-$$
$$76X^5 + 391X^4 + 380X^3-$$
$$410X^2 + 3160X - 580, 1],$$
$$\text{L1}]$$

$$Out[18]= -10 + X + X^2 - \alpha + 4\,X\alpha + 4\alpha^2 - X\alpha^3$$

These procedures allow us to determine, for instance, if a field $\mathbb{Q}(\gamma_1, \ldots, \gamma_n)$ is a subfield of $\mathbb{Q}(\zeta_1, \ldots, \zeta_m)$. This problem is sometimes termed the *subfield immersion problem*. We need only determine whether or not each γ_i is an element of $\mathbb{Q}(\zeta_1, \ldots, \zeta_m)$. Moreover, by detecting whether or not each ζ_i is an element of $\mathbb{Q}(\gamma_1, \ldots, \gamma_n)$, we may determine if two such fields are equal.

21.8. Reduced Forms over Subfields

Given δ in $\mathbb{Q}(\alpha_1, \ldots, \alpha_n)$, we may wish to express δ in reduced form in α over *another* field $\mathbb{Q}(\epsilon_1, \ldots, \epsilon_m)$, where $\mathbb{Q}(\epsilon_1, \ldots, \epsilon_m) \subsetneq \mathbb{Q}(\alpha_1, \ldots, \alpha_n)$, in order that the coefficients of α in the sum be allowed to take values in $\mathbb{Q}(\epsilon_1, \ldots, \epsilon_m)$. To derive this reduced form in AlgFields, we use FMakeTower to declare a new field over $\mathbb{Q}$ by adjoining the ϵ_i successively, followed by FSimplifyE applied to δ over the newly declared field.

We define a field LI as $\mathbb{Q}(\zeta, \nu)$, where $\zeta = 4\epsilon^2 - 20$ and $\nu = \beta + 2$, for α and β in L1; $\mathbb{Q}(\alpha, \beta, \epsilon)$ plays the role of $\mathbb{Q}(\alpha_1, \ldots, \alpha_n)$ above, and ζ and ν play the role of the ϵ_i, so

$\mathbb{Q}(\zeta, v) = \mathbb{Q}(\epsilon_1, \epsilon_2)$. Then `LL` is defined as the extension field of `LI` by adjoining α, β, and ϵ.

```
>   FMakeTower(LL,LI,L1,[zeta,nu],
>       [4*eps^2-20,beta+2]);

----Details of field LI ----

   Algebraic Numbers: [zeta, nu]

   Dimension over Q: 8

   Minimal Polynomials: [-80+zeta^2,

       nu^4-6*nu^3+13*nu^2-

          12*nu-1+1/4*zeta]

   Root Approximations: [

       -8.9442719099991587856,

       -.21464108426685883720]

----
```

```
In[19]:= FMakeTower[LL, LI, L1,

              {ζ, η}, {4ε² − 20, β + 2}]

----Details of field LI ----

   Algebraic Numbers: {ζ, η}

   Dimension over Q: 8

   Minimal Polynomials: {-80 + ζ^2,

       -1 + ζ/4 - 12 * η +

          13 * η^2 - 6 * η^3 + η^4}

   Root Approximations: {

       -8.9442719099991587856,

       -0.21464108426685883720}

----
```

```
----Details of field LL ----

   Algebraic Numbers: [zeta, nu, alpha,
       beta, eps]

   Dimension over Q: 8

   Minimal Polynomials: [-80+zeta^2,
       nu^4-6*nu^3+13*nu^2-12*nu-1+1/4*zeta,
       alpha+2-3*nu+nu^2,
       beta+2-nu,
       eps+(-1/2-1/8*zeta)*nu^2+
          (3/2+3/8*zeta)*nu-1-1/4*zeta]

   Root Approximations: [
       -8.9442719099991587856,

       -.21464108426685883720,

       -2.6899940478558293079,

       -2.2146410842668588372,

       -1.6625077511098137144]

----
```

```
----Details of field LL ----

   Algebraic Numbers: {ζ, η, α, β, ε}

   Dimension over Q: 8

   Minimal Polynomials: {-80 + ζ^2,
       -1 + ζ/4 - 12 * η +
          13 * η^2 - 6 * η^3 + η^4,
       2 + α - 3 * η + η^2,
       2 + β - η,
       -1 + ε - ζ/4 + (3/2 + (3 * ζ)/8) *
          η + (-1/2 - ζ/8) * η^2}

   Root Approximations: {
       -8.9442719099991587856,

       -0.21464108426685883720,

       -2.6899940478558293078,

       -2.2146410842668588372,

       -1.6625077511098137144}

----
```

Then we use `FSimplifyE` to reduce $\epsilon^3 + 7$ over `LL`. Note that since `LI = LL`, $\epsilon^3 + 7$ is expressed in reduced form over `LI` itself.

`>   FSimplifyE(eps^3+7,LL);`	`In[20]:= FSimplifyE[`$\epsilon^3 + 7$`, LL]`
$$(5 + \frac{3\zeta}{4})v^2 + (-\frac{9\zeta}{4} - 15)v + 17 + \frac{3\zeta}{2}$$	$$Out[20]= 17 + \frac{3\zeta}{2} + \left(-15 - \frac{9\zeta}{4}\right)\eta + \left(5 + \frac{3\zeta}{4}\right)\eta^2$$

21.9. Determining and Applying Automorphisms of Fields

The function `FMapIsIsoQ` uses the results of this chapter to determine if an automorphism of a field $K = \mathbb{Q}(\alpha_1, \ldots, \alpha_n)$ exists that sends generating algebraic numbers α_i to counterparts α_j. `FMapIsIsoQ` may also be used to determine if an isomorphism exists from a subfield of the form $\mathbb{Q}(\alpha_1, \ldots, \alpha_i)$ of a field K to another subfield of K, sending each α_i to some α_j.[1]

The syntax of `FMapIsIsoQ` requires first the field, optionally followed by the base field if it is not $\mathbb{Q}$, followed by a list. The first element of this list is itself a list of the generating algebraic numbers α_i on which the map is defined. The second element of this list is also a list, a list of the corresponding images, in order, of those algebraic numbers on which the map is defined. If all of the generating algebraic numbers are used, they may appear in any order in the first sublist; otherwise, they must appear in order as they are declared in the field.

We ask if there exists an isomorphism f of $\mathbb{Q}(r1, r2, r3)$ to $\mathbb{Q}(r1, r3, r5)$, both subfields of `KS`, such that $f(r1) = r1$, $f(r2) = r5$, and $f(r3) = r3$.

`>   FMapIsIsoQ(KS,[[r1,r2,r3],[r1,r5,r3]]);`	`In[21]:= FMapIsIsoQ[KS, {{r1, r2, r3}, {r1, r5, r3}}]`
false	`Out[21]= False`

Such an f does not exist. We check instead if there exists an automorphism f of `KS` over $\mathbb{Q}$ satisfying $f(r1) = f(r1)$, $f(r2) = r5$, $f(r3) = r4$, $f(r4) = r3$, and $f(r5) = r2$.

---

[1] Note that if a map is defined by sending elements *not* in a generating list of K to other elements, or if elements in a generating list are sent to other elements not in the list, we may use `FMakeTower` or `FDeclareExtensionField` to add those additional algebraic numbers to the appropriate lists.

```
>   FMapIsIsoQ(KS,[[r1,r2,r3,r4,r5],
>       [r1,r5,r4,r3,r2]]);
```
$$\mathit{true}$$

In[22]:= FMapIsIsoQ[KS, {{r1, r2, r3, r4, r5},

{r1, r3, r2, r5, r4}}]

Out[22]= True

Now, knowing that f is an automorphism of KS, we apply it to the algebraic number $r1^2 + r1^6 r2 - r5$ using the function FMap.

```
>   FMap(r1^2+r1^6*r2-r5,KS,
>       [[r1,r2,r3,r4,r5],[r1,r5,r4,r3,r2]]);
```
$$-r1^4\,r2^3 - 2\,r2^2 + (-1 - 2\,r1)\,r2 - r1^2$$

In[23]:= FMap[$r1^2 + r1^6 * r2 - r3$, KS,

{{r1, r2, r3, r4, r5}, {r1, r3, r2, r5, r4}}]

Out[23]= $-r1^2 + (-1 - 2\,r1)\,r2 - 2\,r2^2 - r1^4\,r2^3$

We further check if there exists an automorphism g of KS over $\mathbb{Q}$ satisfying $f(r1) = f(r1)$, $f(r2) = r5$, $f(r3) = r3$, $f(r4) = r2$, and $f(r5) = r5$.

```
>   FMapIsIsoQ(KS,
>       [[r1,r2,r3,r4,r5],[r1,r5,r3,r2,r4]]);
```
$$\mathit{false}$$

In[24]:= FMapIsIsoQ[KS,

{{r1, r2, r3, r4, r5}, {r1, r3, r5, r4, r2}}]

Out[24]= False

We determine whether there exists an automorphism h of K2S over K2 satisfying $f(\theta1) = \theta1$, $f(\theta2) = \theta3$, and $f(\theta3) = \theta2$.

```
>   FMapIsIsoQ(K2S,K2,
>       [[th1,th2,th3],[th1,th3,th2]]);
```
$$\mathit{false}$$

In[25]:= FMapIsIsoQ[K2S, K2,

{{$\theta1, \theta2, \theta3$}, {$\theta1, \theta3, \theta2$}}]

Out[25]= False

22. Exercise Set 2

A.

22.1 Following the necessary and sufficient conditions of the Extension of Automorphism to a Multiply Generated Field Theorem (18.3), find all automorphisms of $\mathbb{Q}(\sqrt{2})$. (Hint: There are two.)

22.2 Following the necessary and sufficient conditions of the Extension of Automorphism to a Multiply Generated Field Theorem (18.3), find all automorphisms of $\mathbb{Q}(\sqrt{2}, \sqrt{3})$. (Hint: There are four.)

22.3 Following the necessary and sufficient conditions of the Extension of Automorphism to a Multiply Generated Field Theorem (18.3), find all automorphisms of $\mathbb{Q}(\sqrt{2}, \sqrt{3}, \sqrt{5})$.

22.4 Find the number of automorphisms of $\mathbb{Q}(\sqrt[5]{5})$. Why is this number independent of the choice of $\sqrt[5]{5}$?

22.5 Let K be the splitting field of $X^6 - 1$ over $\mathbb{Q}$. Prove that $K = \mathbb{Q}(\sqrt{-3})$.

22.6 Let K be the splitting field of $X^6 + 1$ over $\mathbb{Q}$. Prove that $K = \mathbb{Q}(\sqrt{3}, \sqrt{-1})$.

22.7 Find all automorphisms of the splitting field of $X^6 + 3X^4 - 1$ over $\mathbb{Q}$. Prove that the splitting field is identical to that of $X^4 + 6X^2 - 8X + 9$. (Hint: Factor the second polynomial over the splitting field of the first and deduce.)

22.8 For each of the following polynomials, find the dimension of the splitting field over $\mathbb{Q}$ and a single generating algebraic number α such that the splitting field is $\mathbb{Q}(\alpha)$:

(1) $X^4 + 1$;

(2) $X^4 + 4$;

(3) $(X^4 - 1)(X^4 + 1)$;

(4) $X^5 - 2$;

(5) $X^5 + X^3 + X^2 + 1$.

B.

22.9 Let $\alpha_1 \approx -2.34$, $\alpha_2 \approx 2.34$, $\alpha_3 \approx -1.21i$, and $\alpha_4 \approx 1.21i$ be the four roots of $X^4 - 4X^2 - 8$. Prove that $\mathbb{Q}(\alpha_1, \alpha_2)$ is not isomorphic to $\mathbb{Q}(\alpha_1, \alpha_3)$.

22.10 Let α_i be as in the preceding exercise. Find

(1) all automorphisms of $\mathbb{Q}(\alpha_i)$ for $1 \le i \le 4$;

(2) all automorphisms of $\mathbb{Q}(\alpha_i, \alpha_j)$ for $1 \le i < j \le 4$;

(3) all automorphisms of $\mathbb{Q}(\alpha_i, \alpha_j, \alpha_k)$ for $1 \le i < j < k \le 4$; and

(4) all automorphisms of $\mathbb{Q}(\alpha_1, \alpha_2, \alpha_3, \alpha_4)$.

22.11 Let $\alpha_1 \approx -3.05$, $\alpha_2 \approx 1.36$, and $\alpha_3 \approx 1.69$ be the three roots of $X^3 - 7X + 7$. Prove that $\mathbb{Q}(\alpha_1, \alpha_2)$ is not only isomorphic but also identical to $\mathbb{Q}(\alpha_1, \alpha_3)$.

22.12 Let α_i be as in the preceding exercise. Find

(1) all isomorphisms among the $\mathbb{Q}(\alpha_i)$ for $1 \le i \le 3$;

(2) all isomorphisms among the $\mathbb{Q}(\alpha_i, \alpha_j)$ for $1 \le i < j \le 3$; and

(3) all isomorphisms of $\mathbb{Q}(\alpha_1, \alpha_2, \alpha_3)$.

22.13 Show that $i\sqrt{3}$ and $1 + i\sqrt{3}$ are roots of $f(X) = X^4 - 2X^3 + 7X^2 - 6X + 12$. Let K be the splitting field of f over $\mathbb{Q}$. Determine all automorphisms of K.

22.14 Determine all automorphisms of the splitting fields in Exercise 22.8.

22.15 Give an example of two *distinct* field extensions $\mathbb{Q}(\alpha)$ and $\mathbb{Q}(\beta)$ of $\mathbb{Q}$ for which $[\mathbb{Q}(\alpha, \beta) : \mathbb{Q}] \neq [\mathbb{Q}(\alpha) : \mathbb{Q}][\mathbb{Q}(\beta) : \mathbb{Q}]$.

22.16 Determine whether or not $\mathbb{Q}(\sqrt[4]{-4})$ is isomorphic to $\mathbb{Q}(\sqrt{5 + \sqrt{24}})$.

22.17 Determine whether or not $\mathbb{Q}(\sqrt[3]{2}\omega)$ is isomorphic to $\mathbb{Q}(\sqrt[3]{2}\omega^2)$, where ω is a nonreal third root of 1, such as $e^{2\pi i/3}$. Determine if the two fields are equal.

CHAPTER FIVE

The Galois Correspondence

We are now ready to introduce one of the most elegant results in algebra, the Galois correspondence. This correspondence gives us a framework for understanding the relationships between the structure of a splitting field over a field K and the structure of its group of automorphisms over K. Once we establish this correspondence, we go on to study in some detail how the group provides some distinguishing characteristics of the various conjugates over K of an element of the splitting field.

23. Normal Field Extensions and Splitting Fields

The property of splitting fields that encapsulates much of the information necessary for proving steps in the Galois correspondence is that of *normality*. In this section, we introduce the property and explore its connection with splitting fields.

Definition 23.1 (Normal Field Extension in $\mathbb{C}$). Let K be a subfield of $\mathbb{C}$. An algebraic field extension L/K is *normal* if every polynomial $f \in K[X]$ that is irreducible over K and that has at least one root in L contains $n = \deg(f)$ roots in L.

Note that since the roots of an irreducible polynomial in $K[X]$ are distinct (Theorem 11.3), this definition is equivalent to

Definition 23.2 (Normal Field Extension). An algebraic field extension L/K is *normal* if every polynomial $p \in K[X]$ that is irreducible over K and that has at least one root in L factors into linear terms over L.

Clearly a splitting field of an irreducible polynomial $p \in K[X]$ satisfies the condition for the particular polynomial p, but the interest of the property lies in whether the condition is satisfied for *all* irreducible polynomials $q \in K[X]$: each such polynomial must have either *none of its roots* in L or *all of its roots* in L.

Theorem 23.3. *Let K be a subfield of $\mathbb{C}$. Let $p \in K[X]$ be a polynomial and L the splitting field of p over K. Then L/K is a normal field extension.*

Proof. This is a special case of the generalization of Theorem 19.1 in Exercise 20.18.   □

Next, we show that the splitting fields over a subfield K of $\mathbb{C}$ are precisely the normal field extensions of K that are finite.

Theorem 23.4. *Let K be a subfield of $\mathbb{C}$. Then L/K is a splitting field extension if and only if it is finite and normal.*

Proof. A splitting field extension is normal by the previous theorem (23.3) and is finite because the dimension of a splitting field extension is bounded by $n!$, where n is the degree of the polynomial (Theorem 16.7).

Now assume L/K is finite and normal. Then because finite is equivalent to simple and algebraic (by the generalization of Theorem 19.7 in Exercise 20.20), L/K is simple and algebraic. Hence $L = K(\alpha)$ for an algebraic element α. Consider $f = m_{\alpha,K}$. One root of f, namely α, clearly lies in L. By the definition of normal, all of the roots of f lie in L. Hence $L = K(\alpha)$ is the splitting field of f over K.   □

Following tradition, one may then make the following definition:

Definition 23.5 (Galois Extension in $\mathbb{C}$). An extension K/F of subfields of $\mathbb{C}$ is a *Galois extension* if K/F is a finite, normal extension.

We will not use the term Galois extension here, preferring to emphasize the idea of splitting fields. We reserve the term for use in section 35 in the definition of a Galois extension over arbitrary fields.

24. The Galois Group

24.1. Definition and Action

Definition 24.1 (Galois Group of a Splitting Field, of a Polynomial). Let K be a subfield of $\mathbb{C}$ and L a splitting field over K. The *Galois group* of L/K is the set of automorphisms g of L over K, under the operation of function composition. We denote the Galois group of L/K as $\mathrm{Gal}(L/K)$:

$$\mathrm{Gal}(L/K) = \left\{ g \in \mathrm{Aut}(L) \;\middle|\; g(k) = k, \ \forall k \in K \right\}.$$

If L is the splitting field over K of a polynomial $p \in K[X]$, we say that the *Galois group of p over K* is $\mathrm{Gal}(p, K) = \mathrm{Gal}(L/K)$. If K is the field of rational numbers, we say that $\mathrm{Gal}(p) = \mathrm{Gal}(L/\mathbb{Q})$ is *the Galois group of p*.

The fact that $\mathrm{Gal}(L/K)$ is a group is left as an exercise (26.1).

When we consider the Galois group of a particular polynomial p, we understand the action of $\mathrm{Gal}(p) = \mathrm{Gal}(L/K)$ first on the roots of $\alpha_1, \dots, \alpha_n$ of p, and then on all of L, as follows. Every automorphism in the Galois group corresponds uniquely to a permutation of the set $\{\alpha_1, \dots, \alpha_n\}$ (Corollary 18.4). Then, since $L = K(\alpha_1, \dots, \alpha_n)$ is generated by the roots α_i, knowing the permutation of the α_i determines the action on all of L.

It is important to recognize, however, that we may study $\mathrm{Gal}(L/K)$ by its action on certain other finite subsets of L, namely the set of roots $\{\beta_i\}$ of some other polynomial f, all of whose roots lie in L. (Note that the multiply generated field $K(\beta_1, \dots, \beta_m)$ may in fact be a proper subfield of L.)

To specify a language for this discussion, we revisit the definition of conjugate.

Definition 24.2. Let K be a subfield of $\mathbb{C}$, and suppose that $K(\alpha)$ and $K(\beta)$ are algebraic over K. We say that α and β are *conjugate over K* if α and β share the same minimal polynomial over K, that is, if $m_{\alpha,K} = m_{\beta,K}$.

Theorem 24.3 (Action of Galois Group on Conjugates). *Let K be a subfield of $\mathbb{C}$, L a splitting field over K, and W the set of roots $\{\alpha_1, \dots, \alpha_n\}$ in L of an irreducible polynomial $f \in K[X]$. Suppose that W is nonempty.*

Then W is the complete set of roots of f in $\mathbb{C}$, and $\mathrm{Gal}(L/K)$ acts on W, that is, each element of $\mathrm{Gal}(L/K)$ permutes the elements of W.

Proof. Let β be a root of f. Since L is a splitting field over K, L is normal over K (Theorem 23.4). Hence L contains every conjugate of β over K, that is, every root of $m_{\beta,K}$. Since each element $g \in \mathrm{Gal}(L/K)$ leaves K, and hence the coefficients of $m_{\beta,K}$, fixed, and since $m_{\beta,K}(\beta_i) = 0$ for each root β_i, we have the following:

$$0 = g(m_{\beta,K}(\beta_i)) = g(m_{\beta,K})(g(\beta_i)) = m_{\beta,K}(g(\beta_i)). \tag{24.1}$$

Hence, any root β_i of $m_{\beta,K}$ must be sent to another root $g(\beta_i)$ of $m_{\beta,K}$. Now since f is irreducible, then $f = cm_{\beta,K}$ for some nonzero constant $c \in K$, and the roots of f are the roots of $m_{\beta,K}$. Hence we have shown that any element g of $\mathrm{Gal}(L/K)$ permutes the elements of the set $\{\beta_1, \ldots, \beta_m\}$ of β and its conjugates. $\square$

One might think of all the elements of L, then, as grouped into "boxes" of elements, where all of the elements in a given box are roots of some particular polynomial in $K[X]$ that is irreducible over K. The Galois group of L/K, then, acts on each box. (In section 27.3, we will say more about these permutation actions: we will show that the permutation actions are *transitive*.)

We have already seen several examples of Galois groups, since in section 17, we computed the set of all automorphisms of some normal extensions. For instance, in Example 17.1 we determined the Galois group of $X^3 + 3X + 1$, which has six elements and contains an automorphism corresponding to each permutation of the three roots of $X^3 + 3X + 1$. Hence $\mathrm{Gal}(X^3 + 3X + 1)$ is isomorphic to S_3. In Example 17.2, we determined the Galois group of $X^3 - 3X - 1$: having only three elements, $\mathrm{Gal}(X^3 - 3X - 1)$ is isomorphic to $\mathbb{Z}/3\mathbb{Z}$. Additional examples may be found by working some of the exercises in Chapter 4, which essentially ask for the Galois group of other polynomials in $\mathbb{Q}[X]$. At the end of this section, we will return to our three cubic examples (17.1–17.3) in detail, as Examples 24.14–24.16.

It is an open problem to determine whether, given any finite group G, there exists a polynomial $p \in \mathbb{Q}[X]$ such that $\mathrm{Gal}(p) \cong G$. This problem is known as the *inverse Galois problem*.

24.2. The Order of the Galois Group Is the Dimension of the Field Extension

We begin by establishing that the number of automorphisms of a splitting field extension is equal to the dimension of the field extension. We do so by considering the elements of a particular "box" as before.

Theorem 24.4. *Let K be a subfield of $\mathbb{C}$ and L a splitting field over K. Then the number of automorphisms of L over K is equal to $[L : K]$. Hence $|\mathrm{Gal}(L/K)| = [L : K]$.*

Proof. We first show that L/K is the splitting field extension of a polynomial in $K[X]$ that is irreducible over K. Since splitting field extensions are finite and normal (Theorem 23.4) and finite extensions are simple (from the generalization of the Finite = Simple and Algebraic Theorem, in Exercise 20.20), there exists some $\gamma \in L$ such that $L = K(\gamma)$ *and L contains all of the roots of* $m_{\gamma,K}$. Therefore L is the splitting field of $m_{\gamma,K}$ over K.

Now consider the set of automorphisms of L over K. By Theorem 18.6 (take $K' = K$, $\tau = id$, $\alpha = \gamma$, and $\alpha' = \gamma'$), for each root γ' of $m_{\gamma,K}$ there exists an isomorphism from $L = K(\gamma)$ to $K(\gamma')$ over K sending γ to γ'. Observing that L is the splitting field of $m_{\gamma,K}$ over K and hence over $K(\gamma)$ and $K(\gamma')$, we may use Theorem 18.7 to extend these isomorphisms to automorphisms of L. (Alternatively, examining degrees, we find that $L = K(\gamma')$ for any root γ' of $m_{\gamma,K}$ and so the isomorphism $K(\gamma) \to K(\gamma')$ is already an automorphism of L.)

Hence there exist at least $\deg(m_{\gamma,K})$ automorphisms of L over K. But by Corollary 18.4, each automorphism g of L over K sends γ to some root $g(\gamma) = \gamma'$ of $m_{\gamma,K}$, and since $L = K(\gamma)$, g is completely determined by γ'. Hence there are no more than $\deg(m_{\gamma,K})$ automorphisms of L over K, which gives us that the order of $\mathrm{Gal}(L/K)$ is equal to $\deg(m_{\gamma,K})$. Since the degree of $m_{\gamma,K}$ is the dimension $[L : K]$ of the extension (Theorem 8.6), we are done. $\square$

24.3. Subfields Correspond to Subgroups and Vice Versa

One of the most fundamental relationships between a splitting field extension L/K and its Galois group is the correspondence between subfield extensions and subgroups. Elegant and powerful, this correspondence provides a remarkable foundation for many later results.

In this section, we associate to every subfield N of L with $K \subset N \subset L$ a subgroup $\mathrm{Gal}(L/N)$ of $\mathrm{Gal}(L/K)$, and to every subgroup H of $\mathrm{Gal}(L/K)$, we associate a subfield $\mathrm{Fix}(H)$ such that $K \subset \mathrm{Fix}(H) \subset L$. In section 24.4, we show that these two associations are inverses of each other and hence that this correspondence between subgroups and subfields is one-to-one.

Definition 24.5 (Fixed Field). Let K be a subfield of $\mathbb{C}$ and L a splitting field over K with Galois group $\mathrm{Gal}(L/K)$. The *fixed field* $\mathrm{Fix}(H)$ *corresponding to a subgroup H of*

Gal(L/K) is the set of elements

$$\mathrm{Fix}(H) = \Big\{ l \in L \ \Big| \ g(l) = l, \ \forall g \in H \Big\}.$$

Definition 24.6 (Intermediate Field). An *intermediate field* N of a field extension L/K is a field N with $K \subset N \subset L$.

Definition 24.7 (Subgroup Corresponding to Intermediate Field). Let K be a subfield of $\mathbb{C}$ and L a splitting field over K with Galois group Gal(L/K). Let N be an intermediate field of L/K. The *subgroup $S(N)$ of* Gal(L/K) *corresponding to* N is the set of elements

$$S(N) = \Big\{ g \in \mathrm{Gal}(L/K) \ \Big| \ g(n) = n, \ \forall n \in N \Big\}.$$

We leave as exercises showing that $S(N)$ is a group (26.1) and that Fix(H) is a field (26.2). Now we prove two propositions on subgroups and intermediate fields.

Proposition 24.8. *Let K be a subfield of $\mathbb{C}$ and L a splitting field over K. If N is an intermediate field of L/K, then L is a splitting field over N.*

Proof. By definition, $L = K(\alpha_1, \ldots, \alpha_n)$, where the α_i are the roots of some polynomial $p \in K[X]$. But then $L = N(\alpha_1, \ldots, \alpha_n)$ where the α_i are the roots of the polynomial $p \in K[X] \subset N[X]$. (Note that even if the polynomial p is irreducible over K, it might not be irreducible over N.) Hence L is a splitting field over N. □

Proposition 24.9. *Let K be a subfield of $\mathbb{C}$, and let L be a splitting field over K.*

- *For each intermediate field N of L/K, the subgroup $S(N)$ of Gal(L/K) is Gal(L/N).*
- *For each subgroup H of Gal(L/K), the fixed field Fix(H) is an intermediate field of L/K.*

Proof. L is a splitting field over any intermediate field of L/K (Proposition 24.8), hence L is a splitting field over N. Therefore there exists a Galois group Gal(L/N). Each element of Gal(L/N) is an automorphism of L that leaves N elementwise fixed, hence each element of Gal(L/N) is an element of $S(N) \subset$ Gal(L/K). Now any element of $S(N)$ is an automorphism of L that leaves N elementwise fixed, hence is an element of Gal(L/N). Therefore Gal(L/N) = $S(N)$.

Now let Fix(H) be the set of elements of L that are left fixed by all of the elements of H. By Exercise 26.2, Fix(H) is a field under the operations inherited from L. Hence Fix(H) is a subfield of L. Since K is elementwise fixed by every element of $H \subset$ Gal(L/K), $K \subset$ Fix(H) and we are done. □

24.4. Subgroups Correspond to Subfields in a One-to-One Fashion

It may seem that the two associations, a subgroup for each intermediate field and an intermediate field for each subgroup, are clearly inverses of each other. That they are inverses is in fact the fundamental correspondence in Galois theory, but it is not a direct consequence of the definitions. From the definitions, we do obtain directly that if N is an intermediate field, then $N \subset \text{Fix}(\text{Gal}(L/N))$. Similarly, if H is a subgroup of $\text{Gal}(L/K)$, then $H \subset \text{Gal}(L/\text{Fix}(H))$. However, making these inclusions into equalities requires some argument, which is the content of the following theorem.

Theorem 24.10 (Galois Correspondence I). *Let K be a subfield of $\mathbb{C}$, and let L be a splitting field over K. Then subgroups of $\text{Gal}(L/K)$ and intermediate fields of L/K correspond in a one-to-one fashion. More precisely, the two associations $N \mapsto S(N)$ and $H \mapsto \text{Fix}(H)$ are inverses of each other.*

Proof. Let $N' = \text{Fix}(\text{Gal}(L/N))$; we first show that $N = N'$. By the definition of fixed field, $N \subset N'$. Now the order of a Galois group is equal to the degree of the field extension (Theorem 24.4), so $[L : N] = |\text{Gal}(L/N)|$ and $[L : N'] = |\text{Gal}(L/N')|$. By the definition of N', every automorphism of L fixing N elementwise fixes N' elementwise, and, on the other hand, every automorphism of L fixing N' elementwise fixes N elementwise, since $N \subset N'$. Therefore the subgroups $\text{Gal}(L/N)$ and $\text{Gal}(L/N')$ of $\text{Gal}(L/K)$ are equal. Hence $[L : N] = [L : N']$, and by the Tower Law (Exercise 13.8), $[L : N] = [L : N'][N' : N]$, so $[N' : N] = 1$ and $N = N'$.

In order to show that $\text{Gal}(L/\text{Fix}(H)) = H$, we do not have at our disposal the technique of the previous paragraph, since if H is not known to be a Galois group of the form $\text{Gal}(L/N)$, then we do not know that $|H| = [L : N]$. However, by the definition of the Galois group, $H \subset \text{Gal}(L/\text{Fix}(H))$, so that $|H| \leq |\text{Gal}(L/\text{Fix}(H))| = [L : \text{Fix}(H)]$. We want to show then that $[L : \text{Fix}(H)] \leq |H|$, and we do so by showing that $[\text{Fix}(H)(\gamma) : \text{Fix}(H)]$ is less than or equal to $|H|$ for any $\gamma \in L$. Since splitting field extensions are finite (Theorem 23.4) and hence simple (Exercise 20.20), there exists some $\gamma \in L$ such that $L = \text{Fix}(H)(\gamma)$, and hence we will be done.

Let $\gamma \in L$ and $m_{\gamma, K}$ be its minimal polynomial over K. By the isomorphism property and the fact that $m_{\gamma, K}$ has coefficients in K, the image of a root of $m_{\gamma, K}$ under an automorphism of L over K must again be a solution to the equation $m_{\gamma, K}(X) = 0$. In particular, each $h \in H$ must send γ to one of the roots of $m_{\gamma, K}$. (See equation (24.1).) Let $W = \{h(\gamma) \mid h \in H\} \subset L$ be the set of images of γ under H. This set W has at most $|H|$ elements.

We now find a polynomial in $\text{Fix}(H)[X]$ that has γ as a root and that has degree less than or equal to $|H|$. We do so by taking the monic polynomial that has, as distinct roots, the elements of W:

$$f(X) = \prod_{w \in W}(X - w).$$

The degree of $f(X)$ is less than or equal to $|H|$, since $\deg(f) = |W| \le |H|$. We claim that the coefficients of $f(X)$ lie in $\text{Fix}(H)$. Index the elements of W as $w_1, w_2, \ldots, w_n$. An exercise (26.7) shows that the coefficient of X^i in $f(X)$ is $(-1)^{n-i}$ times the sum of all products of $n - i$ distinct elements of W. Moreover, any permutation of the elements of W in such a sum will leave the sum unchanged. Now applying an element $h' \in H$ to every element in the set $\{h(\gamma) \mid h \in H\}$ gives $\{h'h(\gamma) \mid h \in H\} = \{h(\gamma) \mid h \in H\}$ (why?), so each element of H permutes the elements of W. Therefore each coefficient of $f(X)$ lies in $\text{Fix}(H)$. The proof is complete.   □

Example 24.11 (Subfields, subgroups). Let $K = \mathbb{Q}$, and let $p(X) = (X^2 - 2)(X^2 - 3)$ $\in K[X]$. Let L be the splitting field of p over K; then $L = \mathbb{Q}(\sqrt{2}, \sqrt{3})$. Now since $[L : \mathbb{Q}] = 4$, the Galois group has order 4, and with the methods of the previous chapter, we determine that there are four automorphisms of $L/\mathbb{Q}$: the identity id on L, the automorphism f_1 sending $\sqrt{2}$ to $-\sqrt{2}$ and $\sqrt{3}$ to $\sqrt{3}$, the automorphism f_2 sending $\sqrt{2}$ to $\sqrt{2}$ and $\sqrt{3}$ to $-\sqrt{3}$, and the automorphism f_3 sending $\sqrt{2}$ to $-\sqrt{2}$ and $\sqrt{3}$ to $-\sqrt{3}$.

The Galois group therefore is isomorphic to $\mathbb{Z}/2\mathbb{Z} \oplus \mathbb{Z}/2\mathbb{Z}$, and hence there are three proper subgroups, each of order 2. These subgroups of the Galois group are each generated by one of the nonidentity elements of $\text{Gal}(L/K)$. To what subfields do they correspond?

Consider the subgroup S_1 of $\text{Gal}(L/K)$ generated by f_1; $S_1 = \{id, f_1\}$. We seek the fixed field $\text{Fix}(S_1)$, which we denote by N. Because $S_1 = \text{Gal}(L/\text{Fix}(S_1))$ (Proposition 24.9) and $|\text{Gal}(L/N)| = [L : N]$ (Theorem 24.4), the dimension $[L : N] = |\text{Gal}(L/N)| = |S_1| = 2$. Then, by the Tower Law (Exercise 13.8), $[N : K] = 2$. Hence the field N is a quadratic extension of K. While later we will attempt to find a systematic method to determine N, here we make an educated guess. We see that f_1 leaves $\sqrt{3}$ fixed, so $K(\sqrt{3})$ is a subfield of L fixed by S. But $[K(\sqrt{3}) : K] = 2$ and $K(\sqrt{3}) \subset N$, so we must have that $N = K(\sqrt{3})$.

The same method shows that $S_2 = \{id, f_2\}$ fixes the subfield $K(\sqrt{2})$. With a little cleverness, we note that f_3 must fix $\sqrt{2} \cdot \sqrt{3} = \sqrt{6}$, since f_3 negates each of $\sqrt{2}$ and $\sqrt{3}$. Hence $S_3 = \{id, f_3\}$ fixes $K(\sqrt{6})$.

Finally, it is clear that the fixed field of the entire Galois group $\mathrm{Gal}(L/K)$ is K, while the subfield of L fixed by the trivial subgroup $\{id\}$ is L.

As is common in Galois theory, we summarize our results pictorially by drawing two lattices, one of subfields and one of subgroups, with inclusion represented in the upward direction in one lattice and in the downward direction in the other. Then each corresponding subgroup and subfield are in corresponding places in the lattices. More succinctly, we have an *order-reversing* correspondence of lattices:

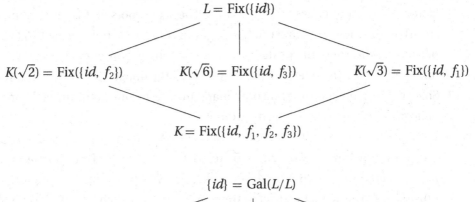

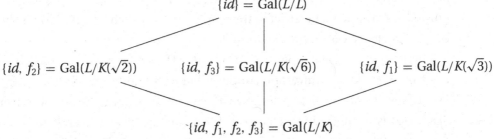

We may find these computations unsurprising. What should be surprising, however, is that, because of the Galois correspondence, *there are no other fields N with $K \subset N \subset L$!*

24.5. Normal Subgroups Correspond to Splitting Fields

The correspondence is still deeper: the fixed fields of normal subgroups of a Galois group $\mathrm{Gal}(L/K)$ are in fact splitting fields over K, and the Galois group $\mathrm{Gal}(L/N)$ of L over an intermediate field N that is a splitting field over K is a normal subgroup of $\mathrm{Gal}(L/K)$.

Theorem 24.12 (Galois Correspondence II). *Let K be a subfield of $\mathbb{C}$, and let L be a splitting field over K. Then normal subgroups of $\mathrm{Gal}(L/K)$ correspond to intermediate fields of L/K that are splitting fields over K.*

Moreover, if N is a splitting field over K that is also an intermediate field of L/K, then $\mathrm{Gal}(N/K) \cong \mathrm{Gal}(L/K)/\mathrm{Gal}(L/N)$, *and* $[N:K] = |\mathrm{Gal}(L/K)|/|\mathrm{Gal}(L/N)|$.

Proof. First suppose that N is both a splitting field over K and an intermediate field of L/K. Then we may consider the Galois group $\mathrm{Gal}(N/K)$ of N over K.

Let $g \in \mathrm{Gal}(L/K)$, and consider the action of g on the field N. By the definition of a splitting field over K, $N = K(\alpha_1, \ldots, \alpha_n)$ where the α_i form a complete set W of roots of a polynomial $f \in K[X]$. Now to use the fact that the Galois group acts on conjugates, we first factor f into irreducibles: $f = f_1^{c_1} \cdots f_k^{c_k}$. The set of roots of f is the disjoint union of the sets W_i of the roots of f_i, and the Galois group acts on the roots of each f_i (Theorem 24.3), permuting the elements of the set W_i. Hence the group permutes the elements of the union W. Since N is generated over K by the α_i, the image of N under g is a subset N' of N. Since g^{-1} is an inverse for g and both maps are one-to-one (as nontrivial homomorphisms of fields), both are isomorphisms from N to N.

Therefore we may define a map $\varphi\colon \mathrm{Gal}(L/K) \to \mathrm{Gal}(N/K)$ that assigns to $g \in \mathrm{Gal}(L/K)$ the automorphism of N obtained by restriction of g to N. That the map is a homomorphism is an exercise (26.3). The kernel $\ker \varphi$ of φ is precisely the subgroup of automorphisms of L over K that restrict to the identity on N, which by definition is the subgroup $\mathrm{Gal}(L/N)$. Hence $\mathrm{Gal}(L/N)$ is a normal subgroup of $\mathrm{Gal}(L/K)$.

Now let H be a normal subgroup of $\mathrm{Gal}(L/K)$. Then H corresponds to a field $\mathrm{Fix}(H) \subset L$. Let $\alpha \in \mathrm{Fix}(H)$ be arbitrary, and suppose that α' is another root of $m_{\alpha,K}$. Splitting field extensions are normal extensions (23.4), so L/K is a normal extension. Hence $\alpha' \in L$. Now $K(\alpha)$ is isomorphic to $K(\alpha')$ under an isomorphism ζ over K that sends α to α' (Theorem 18.6, taking $\tau\colon K \to K' = K$ to be the identity). Then L is a splitting field over each of $K(\alpha)$ and $K(\alpha')$ of the same polynomial $m_{\alpha,K}$ (Proposition 24.8). Hence ζ may then be extended to an automorphism $g\colon L \to L$, which is then an automorphism $g \in \mathrm{Gal}(L/K)$, by Exercise 20.14. Now consider $h \in H$. We have that

$$h(\alpha') = h(g(\alpha)) = (h \circ g)(\alpha) = (g \circ g^{-1} \circ h \circ g)(\alpha).$$

Because H is normal, $g^{-1}hg = h'$ for some element $h' \in H$. Hence

$$h(\alpha') = (g \circ g^{-1} \circ h \circ g)(\alpha) = (g \circ h')(\alpha) = g(h'(\alpha)).$$

Because $h' \in H$ acts as the identity on $\alpha \in \mathrm{Fix}(H)$, we have furthermore that

$$h(\alpha') = g(h'(\alpha)) = g(\alpha) = \alpha'.$$

Therefore, H acts as the identity on the conjugates of each element $\alpha \in \text{Fix}(H)$, which implies that every conjugate of every element $\alpha \in \text{Fix}(H)$ is also in $\text{Fix}(H)$. Hence $\text{Fix}(H)$ is normal over K. Because $\text{Fix}(H)/K$ is finite and normal, $\text{Fix}(H)$ is a splitting field over K (Theorem 23.4).

We have shown that normal subgroups of $\text{Gal}(L/K)$ correspond to intermediate fields of L/K that are splitting fields over K. Now observe that the homomorphism φ defined above maps onto $\text{Gal}(N/K)$ since, given any automorphism of N over K, there exists an automorphism of L extending that automorphism of N, by Theorem 18.7. Therefore $\text{Gal}(N/K)$ is the image of the homomorphism φ, and since φ has kernel $\text{Gal}(L/N)$, the First Isomorphism Theorem for groups tells us that $\text{Gal}(N/K) \cong \text{Gal}(L/K)/\text{Gal}(L/N)$. Since the order of a Galois group is the dimension of the extension (Theorem 24.4), we have the last statement and we are done. $\qquad\square$

Example 24.13 (All subfields normal: $\mathbb{Z}/2\mathbb{Z} \oplus \mathbb{Z}/2\mathbb{Z}$). In Example 24.11, we determined each of the subfields of $L = \mathbb{Q}(\sqrt{2}, \sqrt{3})$ and that $\text{Gal}(L/\mathbb{Q}) \cong \mathbb{Z}/2\mathbb{Z} \oplus \mathbb{Z}/2\mathbb{Z}$. Now since $\mathbb{Z}/2\mathbb{Z} \oplus \mathbb{Z}/2\mathbb{Z}$ is abelian, all of its subgroups are normal. But then, by Theorem 24.12, each of the five subfields $\mathbb{Q}$, $\mathbb{Q}(\sqrt{2})$, $\mathbb{Q}(\sqrt{3})$, $\mathbb{Q}(\sqrt{6})$, and $\mathbb{Q}(\sqrt{2}, \sqrt{3})$ is normal over $\mathbb{Q}$, so they must all be splitting fields over $\mathbb{Q}$ (Theorem 23.4). We may verify this: they are splitting fields, respectively, of X, $X^2 - 2$, $X^2 - 3$, $X^2 - 6$, and $(X^2 - 2)(X^2 - 3)$ over $\mathbb{Q}$.

Example 24.14 (First cubic example; not all subfields normal). Now we return to Example 17.1, where we considered the splitting field L of $X^3 + 3X + 1$ over $\mathbb{Q}$. We determined explicitly that there are precisely six automorphisms of L over $\mathbb{Q}$, one corresponding to each permutation of the three roots α_1, α_2, and α_3. Now the group $\text{Gal}(L/\mathbb{Q})$ acts on the set $\{\alpha_1, \alpha_2, \alpha_3\}$ via these permutations, and we have a permutation representation of $\text{Gal}(L/\mathbb{Q})$ as a subgroup of order 6 of the symmetric group S_3 on three letters. But then $\text{Gal}(L/\mathbb{Q}) \cong S_3$.

Because there exist nonnormal subgroups of S_3, there exist nonnormal intermediate fields of $L/\mathbb{Q}$. Consider the subgroup T generated by the automorphism that exchanges α_1 and α_2 but leaves α_3 fixed. This automorphism is of order 2 in the Galois group and hence generates a subgroup of order 2. Therefore $[L : N] = 2$, where $N = \text{Fix}(T)$; hence by the Tower Law (Exercise 13.8), $[N : \mathbb{Q}] = 3$. One checks that this subgroup T is not normal (since $\text{Gal}(L/\mathbb{Q}) \cong S_3$, show that $\{(), (12)\}$ is not a normal subgroup of S_3). We guess the fixed field to be $\mathbb{Q}(\alpha_3)$, and since $\mathbb{Q}(\alpha_3) \subset \text{Fix}(T)$ and $[\mathbb{Q}(\alpha_3) : \mathbb{Q}] = 3$, we find we are correct. We know then that $\mathbb{Q}(\alpha_3)$ is not normal over $\mathbb{Q}$ by the Galois correspondence

(Theorem 24.12), and we may verify this fact with some computation: α_3 is a root of $X^3 + 3X + 1$, irreducible over $\mathbb{Q}$, but $\mathbb{Q}(\alpha_3)$ does not contain α_1 and α_2, the other roots of the polynomial.

What are the other intermediate fields of $L/\mathbb{Q}$, and which are splitting fields? To aid in the determination, we give the subgroup lattice of S_3. We use the notation $\langle(123)\rangle$ to denote the cyclic subgroup generated by the 3-cycle (123).

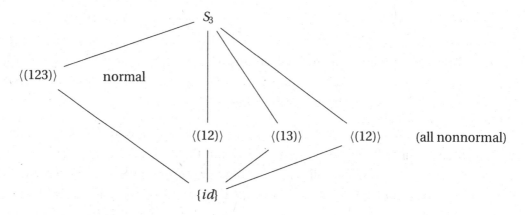

This time, we represent inclusion of subgroups in the upward direction and inclusion of subfields in the downward direction. You should be able to guess every subfield except perhaps one: the fixed field of $\langle(123)\rangle$. This peculiar type of subfield will be the point of departure for the next sections.

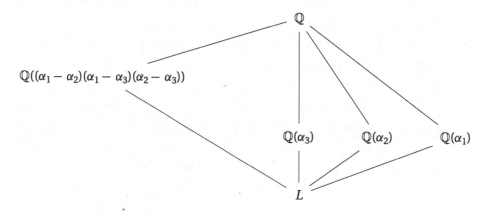

Example 24.15 (Second cubic example). Consider the splitting field L of $X^3 - 3X - 1$ over $\mathbb{Q}$, as in Example 17.2. By the methods of the previous example, the Galois group $\mathrm{Gal}(L/\mathbb{Q})$

is of order 3, isomorphic to the subgroup of S_3 consisting of the three permutations (), (123), and (132). Hence Gal($L/\mathbb{Q}$) is a cyclic group of order 3. Draw the associated subgroup and subfield lattices. What can you deduce?

Example 24.16 (Third cubic example). Finally, consider the splitting field L of the cubic $X^3 - 2$ over $\mathbb{Q}$, as in Example 17.3. We determined that there are precisely six automorphisms of L over $\mathbb{Q}$, one corresponding to each permutation of the three roots α_1, α_2, and α_3. Hence Gal($L/\mathbb{Q}$) $\cong S_3$, just as in Example 24.14. The subgroup diagram of S_3 will, of course, be the same, and the subfield diagram, with inclusions in the upward direction, is then the following:

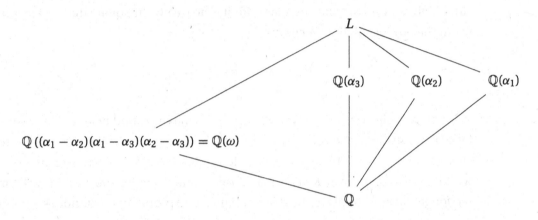

(Here ω denotes α_2/α_1; it is nontrivial third root of unity.)

25. Invariant Polynomials, Galois Resolvents, and the Discriminant

Now that we have established the Galois correspondence, we investigate more closely how the Galois group determines algebraic relationships among the roots of a polynomial. In doing so, we will, in turn, develop additional tools for the determination of the Galois group.

Our point of departure is a curious arithmetic expression in the roots of a polynomial called the *discriminant*. Suppose $p \in K[X]$ is a polynomial with roots $\alpha_1, \ldots, \alpha_n$. The discriminant is the square of a product of differences of roots:

$$\text{disc}(p) = \left(\prod_{1 \leq i < j \leq n} (\alpha_i - \alpha_j) \right)^2 .$$

One useful property of the discriminant is that $\text{disc}(p) = 0$ if and only if the polynomial has roots of multiplicity greater than one. However, the discriminant will be even more useful to us when it is nonzero.

Taking another look, it may seem odd to specify that the differences are always taken so that the root with the higher index is always subtracted from the root with the lower index, since when the product is squared, it makes no difference. But this specification already gives a hint of what lies ahead, for it is not really the square that we are interested in but the product of differences itself:

$$\delta = \prod_{1 \leq i < j \leq n} (\alpha_i - \alpha_j)$$

What we really want to know is whether δ lies in the ground field K. The short answer for why we instead take the square, $\text{disc}(p)$, is that (a) it is easier to calculate from the coefficients of p, and (b) we can easily determine if $\delta \in K$ by factoring the polynomial $X^2 - \text{disc}(p) \in K[X]$ over K. The goal of this section is to lay the groundwork for a better, longer answer, to place the discriminant in its proper context, not as some special element of the ground field K, but as a coefficient of a particular resolvent polynomial $X^2 - \text{disc}(p)$.

Resolvent polynomials, called *Galois resolvents*, are powerful tools for distinguishing the roots of a polynomial, and, as well, for determining the Galois group. If a polynomial p of degree n has Galois group $\text{Gal}(p) \cong S_n$, we surmise that the roots are, relatively speaking, indistinguishable, for we have automorphisms of the splitting field permuting the roots in any way we please. The smaller the Galois group (among subgroups of S_n), the more distinguished the roots must be, and the factorization of resolvent polynomials helps us detect these distinctions.

For now we introduce Galois resolvent polynomials for a given polynomial p, and then we explore how we may calculate the polynomials from the coefficients of p. At the end of the section, we give a proof of the following proposition, which gives a taste of what is to come.

Proposition 25.1. *Let K be a subfield of $\mathbb{C}$. Let $p \in K[X]$ be a polynomial of degree n with distinct roots, and $\alpha_1, \ldots, \alpha_n$ the roots of p. Set $G = \mathrm{Gal}(p, K)$, identified with a subgroup of S_n via its action of the roots $\alpha_1, \ldots, \alpha_n$.*

Then $\mathrm{disc}(p) \in K$, and the following are equivalent:

(1) $\mathrm{disc}(p)$ is the square of an element in K;

(2) $X^2 - \mathrm{disc}(p)$ has a linear factor over K;

(3) $G \subset A_n$.

25.1. Invariant Polynomials and Galois Resolvents

We begin by considering polynomials in several indeterminates permuted by a subgroup of the symmetric group, as well as the orbits and stabilizers of these polynomials.

Suppose that $t = t(X_1, \ldots, X_n) \in K[X_1, \ldots, X_n]$. Consider the action of S_n on $K[X_1, \ldots, X_n]$ by the rule

$$g(t) = t(X_{g(1)}, X_{g(2)}, \ldots, X_{g(n)}), \qquad g \in S_n.$$

Observe that elements of S_n may send t to different polynomials. For example, if $t = X_1 X_2 + X_3$ and $g = (123) \in S_3$, then $g(t) = X_2 X_3 + X_1$.

We say that the *stabilizer* of t is the subgroup $H \subset S_n$ leaving t fixed. In other words, H is precisely the subset of elements $h \in S_n$ such that $t(X_{h(1)}, X_{h(2)}, \ldots, X_{h(n)}) = t(X_1, \ldots, X_n)$:

$$H = \left\{ h \in S_n \;\middle|\; t(X_{h(1)}, X_{h(2)}, \ldots, X_{h(n)}) = t(X_1, \ldots, X_n) \right\}.$$

Check that the stabilizer of $t = X_1 X_2 + X_3$ in S_3 is $\{(), (12)\}$.

We say that the *orbit* of t is the set $T \subset K[X_1, \ldots, X_n]$ of all images of t under permutations of the indeterminates by elements of S_n:

$$T = \left\{ g(t) \;\middle|\; g \in S_n \right\}.$$

Check that the orbit of $t = X_1 X_2 + X_3$ under S_3 is $\{X_1 X_2 + X_3, \ X_2 X_3 + X_1, \ X_1 X_3 + X_2\}$.

We have arrived at the definition of a resolvent.

Definition 25.2. Let K be a subfield of $\mathbb{C}$, L the splitting field over K of a polynomial $p \in K[X]$ of degree n with roots $\alpha_1, \ldots, \alpha_n$, and $t \in K[X_1, \ldots, X_n]$ a polynomial with stabilizer H and orbit T. The *Galois resolvent* (over K) corresponding to t and p is the polynomial in $K[X]$ defined as

$$f_{t,p}(X) = \prod_{u \in T} (X - u(\alpha_1, \ldots, \alpha_n)) .$$

We say that $f_{t,p}$ is a *resolvent for H*.

The resolvent is therefore a polynomial that is a product of linear factors, each of which is the difference of X and an element of T, with α_i substituted for X_i for each i. Note first that by identifying T with $\mathrm{orb}_{S_n}(t)$ and H with $\mathrm{stab}_{S_n}(t)$, the Orbit-Stabilizer Theorem (4.6) tells us that $|T| = |S_n|/|H|$, so $\deg(f_{t,p}) = |S_n|/|H|$. (The careful reader will note that $K[X_1, \ldots, X_n]$ is not a finite set upon which S_n acts; however, since S_n is finite, we may think of S_n as permuting a finite set, namely the orbit T of t.)

We must still justify the statement that $f_{t,p}(X) \in K[X]$, however, and we do so with the following theorem.

Theorem 25.3. *Let K be a subfield of $\mathbb{C}$. The Galois resolvents over K lie in $K[X]$: if $p \in K[X]$ is a polynomial of degree n and $t \in K[X_1, \ldots, X_n]$, then $f_{t,p} \in K[X]$.*

Proof. Let the splitting field of p over K be denoted L, and let $\alpha_1, \ldots, \alpha_n$ be the roots of p. Denote by $\varphi : K[X_1, \ldots, X_n] \to L$ the evaluation homomorphism $\varphi(f) = f(\alpha_1, \ldots, \alpha_n)$.

Consider the coefficient c_i of X^i in the Galois resolvent $f_{t,p}(X)$. By Exercise 26.7, c_i is $(-1)^{n-i}$ times the sum of all products of $n - i$ elements $\varphi(u)$, where u is chosen from T. Hence[1]

$$c_i = (-1)^{n-i} \sum_{\substack{U \subset T \\ |U|=n-i}} \prod_{u \in U} \varphi(u). \tag{25.1}$$

Now let $g \in \mathrm{Gal}(L/K)$, which we identify with a subgroup of S_n by virtue of its action on the roots $\alpha_1, \ldots, \alpha_n$. Under this identification, we may write $\alpha_i \mapsto \alpha_{g(i)}$ and consider this same action $i \mapsto g(i)$ on the polynomial ring $K[X_1, \ldots, X_n]$ by sending $X_i \mapsto X_{g(i)}$. We check then that g commutes with the evaluation map φ, as follows. On an arbitrary X_i, $\varphi \circ g$ acts by

$$X_i \mapsto X_{g(i)} \mapsto \alpha_{g(i)},$$

---

[1] Here we follow mathematical tradition in writing below a sum or a product the index set of the sum or the product: a set of objects at which the argument of the sum or product is evaluated. Here we sum over all subsets U of T of cardinality $n - i$, and, for each such subset, we take a product over all the elements u of U.

while $g \circ \varphi$ acts by

$$X_i \mapsto \alpha_i \mapsto \alpha_{g(i)}.$$

Hence $g\varphi = \varphi g$.

Let $u \in T$ be arbitrary. Then $u = s(t)$ for some $s \in S_n$. We calculate:

$$g\varphi(u) = g\varphi(s(t)) = \varphi(g(s(t))) = \varphi((gs)(t)).$$

Hence $g\varphi(u)$ is another element of T evaluated at $\alpha_1, \ldots, \alpha_n$, namely $gs(t)$. Referring to equation (25.1), we see that any element $g \in \mathrm{Gal}(L/K)$ applied to c_i simply permutes the factors in the products and the summands in the sum, leaving the result c_i unchanged (Exercise 26.7). Hence for every $g \in \mathrm{Gal}(L/K)$, $g(c_i) = c_i$, and so $f_{t,p} \in \mathrm{Fix}(\mathrm{Gal}(L/K))[X]$.

Now by the Galois correspondence, $\mathrm{Fix}(\mathrm{Gal}(L/N)) = N$ for an intermediate field N of L/K (Theorem 24.10). Taking $N = K$, we have $\mathrm{Fix}(\mathrm{Gal}(L/K)) = K$ and hence that $f_{t,p} \in K[X]$. □

In the following examples we examine some resolvents.

Example 25.4 (Resolvent for $A_4 \subset S_4$). The polynomial

$$t(X_1, X_2, X_3, X_4) = (X_1 - X_2)(X_1 - X_3)(X_1 - X_4)(X_2 - X_3)(X_2 - X_4)(X_3 - X_4) \quad (25.2)$$

has stabilizer $H = A_4 \subset S_4$, as follows. Any element of S_4 may be written as a product of transpositions (ij) with $i < j$. Each such transposition exchanges $(X_i - X_j)$ with $(X_j - X_i) = -(X_i - X_j)$, exchanges $X_i - X_k$ with $X_j - X_k$ and $X_k - X_j$ with $X_k - X_i$ for $k \notin \{i, j\}$, and leaves all other $X_k - X_l$, $k, l \notin \{i, j\}$ fixed. Except for $X_i - X_j$, if an exchange negates a sign in the product in equation (25.2), the exchange negates the signs of both factors in the pair. Hence a transposition changes the sign of the product. Every product of an even number of transpositions, then, leaves t fixed, while every product of an odd number of transpositions negates t. The subset of elements that leave t fixed is then precisely A_4.

Now the orbit of t is the set T of all images of t under the action of S_4. Since any product of transpositions either leaves t fixed or negates t, there are only two such images, namely t and $-t$, so that the orbit is $T = \{t, -t\}$.

We then deduce that the resolvent polynomial associated to a polynomial p with roots $\alpha_1, \ldots, \alpha_4$ is

$$\big(X - t(\alpha_1, \ldots, \alpha_4)\big)\big(X + t(\alpha_1, \ldots, \alpha_4)\big) = X^2 - t(\alpha_1, \ldots, \alpha_4)^2,$$

or

$$f_{t,p}(X) = X^2 - (\alpha_1 - \alpha_2)^2(\alpha_1 - \alpha_3)^2(\alpha_1 - \alpha_4)^2(\alpha_2 - \alpha_3)^2(\alpha_2 - \alpha_4)^2(\alpha_3 - \alpha_4)^2.$$

The last coefficient is nothing but $-\mathrm{disc}(p)$. In this form, it does not appear that $f_{t,p}(X) \in K[X]$, although the previous theorem tells us so. Later we will see how we can determine the coefficients in K algorithmically, based on the coefficients of p.

Example 25.5 (Resolvent for $V \subset S_4$). The polynomial

$$t(X_1, X_2, X_3, X_4) = X_1 X_2^2 X_3^3 + X_1^2 X_2 X_4^3 + X_1^3 X_3 X_4^2 + X_2^3 X_3^2 X_4$$

has stabilizer $V = \{1, (12)(34), (13)(24), (14)(23)\} \subset S_4$, as follows. By inspection, the elements of V leave t fixed. Now the six cosets of V in S_4 are V, $(12)V$, $(13)V$, $(14)V$, $(123)V$, and $(132)V$, and by inspection, we check that each of (12), (13), (14), (123), and (132) alter t. These calculations reveal an orbit of t that consists of six elements, and hence the resolvent associated to a polynomial p of degree 4 and to t is a polynomial $f_{t,p}$ of degree 6.

The polynomial t from the last example was not chosen arbitrarily. It is a cousin of

$$t_H(X_1, \ldots, X_n) = \sum_{h \in H} \prod_{i=1}^n X_{h(i)}^i = \sum_{h \in H} X_{h(1)} X_{h(2)}^2 \cdots X_{h(n)}^n,$$

which, for an arbitrary subgroup H of S_n, is a polynomial with stabilizer H (Exercise 26.9).

25.2. Symmetric Polynomials and Resolvent Coefficients

Galois resolvents over K are polynomials in $K[X]$ (Theorem 25.3), and thus far we know the coefficients of Galois resolvents only via the definition of a resolvent: as arithmetic combinations of the roots of polynomial $p \in K[X]$. Now we could reduce these expressions to a *normal form* of elements in a multiply generated field (see Exercise 20.5), and doing so would ultimately reduce these coefficients to their equivalent expressions as elements of K. However, working with this normal form requires the knowledge of minimal polynomials of each root α_i of p over fields $K_{i-1} = K(\alpha_1, \ldots, \alpha_{i-1})$. Instead, we develop a procedure to determine the coefficients of Galois resolvents from the coefficients of the original polynomial p, working only over the base field K.

Definition 25.6 (Symmetric Polynomial). Let K be a field. A *symmetric polynomial* in a polynomial ring $K[X_1, \ldots, X_n]$ is a polynomial that is invariant under the action of S_n on the indices of the indeterminates. Equivalently, a symmetric polynomial is a polynomial with stabilizer S_n.

Definition 25.7 (Elementary Symmetric Polynomials). Let K be a field. The n *elementary symmetric polynomials* $\sigma_1, \sigma_2, \ldots, \sigma_n$ of $K[X_1, \ldots, X_n]$ are defined by

$$\sigma_j(X_1, \ldots, X_n) = \sum_{1 \le i_1 \le i_2 \le \cdots \le i_j \le n} X_{i_1} \cdots X_{i_j}, \qquad j \in \{1, 2, \ldots, n\}.$$

We say that σ_j is the *elementary symmetric polynomial of degree j*.

For $n = 3$, for example, $\sigma_1 = X_1 + X_2 + X_3$ and $\sigma_2 = X_1 X_2 + X_1 X_3 + X_2 X_3$. Note that an alternative definition of the jth elementary symmetric polynomial is

$$\sigma_j(X_1, \ldots, X_n) = \sum_{\substack{U \subset \{1,2,\ldots,n\} \\ |U| = j}} \prod_{i \in U} X_i.$$

The symmetric polynomials which are sums of jth powers of the variables (such as $X_1^2 + X_2^2 + X_3^2$ and $X_1^3 + X_2^3 + X_3^3$) are also interesting, and the relations between the elementary symmetric polynomials and these are called *Newton's identities*. For a recent article on the connection, see [35].

Every symmetric polynomial over K may be expressed as a polynomial over K "in the elementary symmetric polynomials"; in other words, it may be expressed as a polynomial in the "indeterminates" σ_j, not only as a polynomial in the indeterminates X_j. This fact is the content of the following theorem.

Theorem 25.8. *Let $f \in K[X_1, \ldots, X_n]$ be a symmetric polynomial. Then there exists a polynomial $q \in K[X_1, \ldots, X_n]$ such that $q(\sigma_1, \ldots, \sigma_n) = f$.*

Note that although the polynomial q is given in terms of variables X_i, we will substitute polynomials $\sigma_i(X_1, \ldots, X_n)$ for each of these X_i, and these substituted polynomials will be in terms of the X_i again. If you are confused by this notation, imagine that $q \in K[Y_1, \ldots, Y_n]$ and perform the substitutions $Y_i = \sigma_i(X_1, \ldots, X_n)$.

Example 25.9. Consider the symmetric polynomial

$$f = X_1 X_2^2 + X_1 X_3^2 + X_2 X_1^2 + X_2 X_3^2 + X_3 X_1^2 + X_3 X_2^2 \in K[X_1, X_2, X_3].$$

By definition, the elementary symmetric polynomials for $n = 3$ are

$$\sigma_1(X_1, X_2, X_3) = X_1 + X_2 + X_3,$$

$$\sigma_2(X_1, X_2, X_3) = X_1 X_2 + X_1 X_3 + X_2 X_3,$$

$$\sigma_3(X_1, X_2, X_3) = X_1 X_2 X_3.$$

Now we check that $f = \sigma_1 \sigma_2 - 3\sigma_3$, hence $q(X_1, X_2, X_3) = X_1 X_2 - 3X_3 \in K[X_1, X_2, X_3]$ is such that $q(\sigma_1, \sigma_2, \sigma_3) = f$. We have then expressed f as a polynomial in the σ_i.

Proof of Theorem 25.8. First we define a partial ordering on the set of monomials

$$\{X_1^{k_1} X_2^{k_2} \cdots X_n^{k_n}\}_{k_j \geq 0}.$$

We write

$$X_1^{k_1} X_2^{k_2} \cdots X_n^{k_n} < X_1^{l_1} X_2^{l_2} \cdots X_n^{l_n}, \qquad k_i, l_i \geq 0$$

if there exists a $j \leq n$ such that $k_i = l_i$ for $i < j$ and $k_j < l_j$. We define $f \leq g$ for two monomials f and g to mean that either $f < g$ or $f = g$. It is an exercise (26.10) to show that if $f \leq f'$ and $g \leq g'$, then $fg \leq f'g'$.

We view f as a sum of K-multiples of monomials of the form $X_1^{k_1} X_2^{k_2} \cdots X_n^{k_n}$; note for instance that a constant term c in f is then $c \cdot 1$, where 1 is the monomial with all $k_i = 0$. Now that we have an ordering on monomials, we prove the theorem by induction on the greatest monomial occurring in the symmetric polynomial f. (We say that a monomial y *occurs* in f if, when written as a sum of monomials, cy appears for some nonzero constant $c \in K$.) Clearly the smallest monomial is 1.

If 1 is the highest monomial in f, then $f = k$ for some constant $k \in K$, and our statement is true for $q(X_1, \ldots, X_n) = k$. Now note that the greatest monomial $y = X_1^{k_1} X_2^{k_2} \cdots X_n^{k_n}$ of a symmetric polynomial f must satisfy $k_1 \geq k_2 \geq \cdots \geq k_n$, as follows. If not, there exists a permutation in S_n that, when applied to the indeterminates $X_1, \ldots, X_n$, sends y into a monomial y' with the desired nonincreasing sequence of exponents. Now since f is a symmetric polynomial, applying this permutation to f yields f; hence the monomial y' occurs in f. But $y' \geq y$, which contradicts the maximality of y. Therefore we need only induct on monomials with a nonincreasing sequence of exponents.

For the inductive step, assume three things: (a) that f is a symmetric polynomial with greatest monomial y, so that ky occurs in f for some nonzero $k \in K$; (b) that y has nonincreasing exponents; and (c) that any symmetric polynomial f' that is a sum of K-multiples of monomials less than y may be expressed as $q'(\sigma_1, \ldots, \sigma_n)$ for some $q' \in K[X_1, \ldots, X_n]$. Then let $y = X_1^{k_1} X_2^{k_2} \cdots X_n^{k_n}$ and consider

$$r = \sigma_1^{k_1-k_2} \sigma_2^{k_2-k_3} \cdots \sigma_{n-1}^{k_{n-1}-k_n} \sigma_n^{k_n}.$$

Since $k_i \geq k_{i+1}$ for $i = 1, \ldots, n-1$, each exponent in the product defining r is nonnegative and so $r \in K[X_1, \ldots, X_n]$. It is an exercise (26.11) to show that the greatest monomial of r is $X_1^{k_1} X_2^{k_2} \cdots X_n^{k_n} = y$. Now let $k \neq 0$ be the coefficient of y in f. Then $f(X_1, \ldots, X_n) - kr$ contains only monomials less than y. By induction, then there exists a polynomial $q' \in K[X_1, \ldots, X_n]$ such that $q'(\sigma_1, \ldots, \sigma_n) = f(X_1, \ldots, X_n) - kr(X_1, \ldots, X_n)$. But then

$$q(X_1, \ldots, X_n) = q'(X_1, \ldots, X_n) + kX_1^{k_1-k_2} X_2^{k_2-k_3} \cdots X_n^{k_n} \in K[X_1, \ldots, X_n]$$

is precisely the polynomial we need.[2]   □

This proof is algorithmic, and we wish to use this method to find the coefficients of a resolvent in terms of the coefficients of p. In particular, given a $t \in K[X_1, \ldots, X_n]$, we seek a formula for $f_{t,p}$ in terms of the *coefficients* of p, not the *roots* of p. We will produce, in fact, a general formula to find $f_{t,p}(X)$ for any p of given degree n. While computing $f_{t,p}$ for general p of a given degree is nontrivial, once we achieve the formula, it will be a simple matter to determine $f_{t,p}$ for any p of that degree.

Fix $t \in K[X_1, \ldots, X_n]$, and let $p \in K[X]$ be a polynomial of degree n. Consider the coefficient d_i of X^i in the Galois resolvent $f_{t,p}(X) = \sum_{i=0}^{\deg(f_{t,p})} d_i X^i$. By Exercise 26.7 and the definition of Galois resolvent, d_i equals $(-1)^{n-i}$ times the sum of all products of $n - i$ distinct elements chosen from the orbit T of t, each evaluated at $\alpha_1, \ldots, \alpha_n$.

However, if we stop before the evaluation at $\alpha_1, \ldots, \alpha_n$, then we may consider a polynomial we call $z_i(X_1, \ldots, X_n)$, defined as $(-1)^{n-i}$ times the sum of all products of $n - i$

---

[2] An alternative approach offering some additional intuition – but assuming a greater generality of the Galois correspondence than we have developed (see section 35) – proceeds as follows. Let L be the field of rational functions $K(X_1, \ldots, X_n)$, that is, the field of quotients of polynomials in $K[X_1, \ldots, X_n]$ by nonzero polynomials in $K[X_1, \ldots, X_n]$. Now consider the subfield $K(\sigma_1, \ldots, \sigma_n)$ generated over K by the elementary symmetric polynomials. Then L is the splitting field of $X^n - (\sigma_1)X^{n-1} + \cdots + (-1)^n(\sigma_n)$, since this polynomial is equal to the product $(X - X_1)(X - X_2) \cdots (X - X_n)$. The Galois group, viewed as a group of permutations of $\{X_1, \ldots, X_n\}$, must be a subgroup of S_n; on the other hand, every permutation in S_n gives a different automorphism of L over $K(\sigma_1, \ldots, \sigma_n)$. Hence $K(X_1, \ldots, X_n)/K(\sigma_1, \ldots, \sigma_n)$ is Galois with group S_n, and $K(\sigma_1, \ldots, \sigma_n)$ is the fixed field of S_n. To perform the final step – to say that every symmetric polynomial is a polynomial in the elementary symmetric functions, that is, that each symmetric polynomial lies not only in $K(\sigma_1, \ldots, \sigma_n)$ but $K[\sigma_1, \ldots, \sigma_n]$ – requires a notion of integrality beyond the scope of this text. However, see [21, §I.3, Ex. 17].

distinct elements chosen from the orbit T:

$$z_i(X_1, \ldots, X_n) = (-1)^{n-i} \sum_{\substack{U \subset T \\ |U|=n-i}} \prod_{u \in U} u(X_1, \ldots, X_n).$$

Since S_n permutes the elements of T, each element of S_n permutes the factors in the products and the summands in the sum, leaving the result, $z_i(X_1, \ldots, X_n)$, fixed. Hence $z_i(X_1, \ldots, X_n)$ is a symmetric polynomial and d_i is the evaluation, at $\alpha_1, \ldots, \alpha_n$, of the symmetric polynomial z_i in $X_1, \ldots, X_n$.

Since z_i is symmetric, $z_i(X_1, \ldots, X_n) = q_i(\sigma_1, \ldots, \sigma_n)$ for a polynomial $q_i \in K[X_1, \ldots, X_n]$ (Theorem 25.8). Now let $\varphi \colon K[X_1, \ldots, X_n] \to K(\alpha_1, \ldots, \alpha_n)$ be the evaluation homomorphism $\varphi(f) = f(\alpha_1, \ldots, \alpha_n)$. We deduce

$$
\begin{aligned}
d_i &= \varphi(z_i) = z_i(\alpha_1, \ldots, \alpha_n) \\
&= q_i(\sigma_1(\alpha_1, \ldots, \alpha_n), \sigma_2(\alpha_1, \ldots, \alpha_n), \ldots, \sigma_n(\alpha_1, \ldots, \alpha_n)) \qquad (25.3) \\
&= q_i(\varphi(\sigma_1), \ldots, \varphi(\sigma_n)), \quad i = 0, \ldots, \deg(f_{t,p}).
\end{aligned}
$$

But by Exercise 26.7, we may determine these $\varphi(\sigma_i) = \sigma_i(\alpha_1, \ldots, \alpha_n)$, as follows. The coefficient c_i of X^i in a monic polynomial $p(X) = \sum_{i=0}^{\deg(p)} c_i X^i \in K[X]$ with roots (taken with multiplicities, if necessary) $\alpha_1, \ldots, \alpha_n$ is

$$c_i = (-1)^{n-i}\sigma_{n-i}(\alpha_1, \ldots, \alpha_n) = (-1)^{n-i}\varphi(\sigma_{n-i}), \qquad i = 0, \ldots, n-1.$$

Knowing the coefficients of the polynomial p, then, we may determine $\varphi(\sigma_i)$:

$$\varphi(\sigma_i) = (-1)^i c_{n-i}, \qquad i = 1, \ldots, n. \qquad (25.4)$$

Now put equations (25.3) and (25.4) together:

$$d_i = q_i(-c_{n-1}, c_{n-2}, -c_{n-3}, \ldots, (-1)^n c_0).$$

We have deduced a procedure verifying the following theorem.

Theorem 25.10. *Let K be a field and $t \in K[X_1, \ldots, X_n]$. Then we may compute the coefficients of $f_{t,p}(X)$ as polynomials in the coefficients c_i of a monic polynomial $p \in K[X]$ of degree n; moreover, for a particular p, we may substitute the particular values of the coefficients c_i into our expression to calculate $f_{t,p}(X)$.*

25.3. The Discriminant

Observe that by generalizing the argument of Example 25.4, for each n, the polynomial

$$\delta_n(X_1, \ldots, X_n) = \prod_{i<j}(X_i - X_j) \in K[X_1, \ldots, X_n]$$

has stabilizer $A_n \subset S_n$, and that for a polynomial $p \in K[X]$ of degree n with roots $\alpha_1, \ldots, \alpha_n$,

$$f_{\delta_n, p}(X) = X^2 - \delta_n(\alpha_1, \ldots, \alpha_n)^2 \in K[X]$$

is a resolvent for A_n associated to δ_n and $p \in K[X]$.

Of course, we have returned to our curious friend, the discriminant:

Definition 25.11 (Discriminant, Discriminant Resolvent). The *discriminant* disc(p) of a nonconstant polynomial $p \in K[X]$ with roots $\alpha_1, \ldots, \alpha_n$ is disc(p) $= (\delta_n(\alpha_1, \ldots, \alpha_n))^2 \in K$.

The *discriminant resolvent* of a nonconstant polynomial $p \in K[X]$ with roots $\alpha_1, \ldots, \alpha_n$ is $f_{\delta_n, p} = X^2 - (\delta_n(\alpha_1, \ldots, \alpha_n))^2 = X^2 - \text{disc}(p) \in K[X]$.

In section 25.2, we determined a procedure for finding a general formula for any resolvent, given a polynomial of fixed degree; we put this procedure into practice in the following example for discriminant resolvents for polynomials of degree two.

Example 25.12 (Discriminant of $aX^2 + bX + c$). Let $p(X) = aX^2 + bX + c \in K[X]$, $a \neq 0$, with roots α_1, α_2. Set $t = \delta_2 = X_1 - X_2$ so that

$$f_{t, p}(X) = X^2 - \text{disc}(p) = X^2 - \delta_2^2(\alpha_1, \alpha_2).$$

We rewrite $\delta_2^2 = (X_1 - X_2)^2$ using the procedure of Theorem 25.8. First we expand δ_2^2 out:

$$(X_1 - X_2)^2 = X_1^2 - 2X_1 X_2 + X_2^2.$$

The greatest monomial is $X_1^2 = X_1^2 X_2^0$, with coefficient 1. Hence we consider

$$\sigma_1^{2-0}\sigma_2^0 = \sigma_1^2 = (X_1 + X_2)^2 = X_1^2 + 2X_1 X_2 + X_2^2.$$

Subtracting σ_1^2 from δ_2^2 produces $-4X_1 X_2$, which has greatest monomial $X_1^1 X_2^1$ with coefficient -4. Hence we further subtract

$$-4\sigma_1^{1-1}\sigma_2^1 = -4\sigma_2 = -4X_1 X_2,$$

leaving nothing left. Hence $\delta_2^2 = \sigma_1^2 - 4\sigma_2$, or $q(\sigma_1, \sigma_2)$ for $q(X_1, X_2) = X_1^2 - 4X_2$.

Now by equation (25.4) and the analysis preceding it, $\text{disc}(p) = q(-c_1, c_0)$, where $X^2 + c_1 X + c_0$ is a *monic* polynomial with roots α_1, α_2. We find such a polynomial by dividing our original polynomial $aX^2 + bX + c$ by a: $X^2 + (b/a)X + (c/a)$. Therefore

$$\text{disc}(p) = q(-b/a,\ c/a) = (b/a)^2 - 4(c/a) = (b^2 - 4ac)/a^2.$$

For monic polynomials of degree 2, then, the discriminant is, up to squares, the familiar formula underneath the square root in the quadratic formula: $a^2 \text{disc}(p) = b^2 - 4ac$.

Now we prove Proposition 25.1.

Proof of Proposition 25.1. By Theorem 25.3, the Galois resolvent $f_{\delta_n, p} \in K[X]$. Since $-\text{disc}(p)$ is the constant term of $f_{\delta_n, p}$, we deduce that $\text{disc}(p) \in K$.

Clearly the first item implies the second. For the reverse, observe that if $X^2 - \text{disc}(p)$ factors into linear terms over K, then it has a root $\alpha \in K$ and $\alpha^2 = \text{disc}(p)$.

Now the stabilizer of δ_n is A_n and any odd element of S_n sends δ_n to its negative (see the argument of Example 25.4). Let $\varphi \colon K[X_1, \ldots, X_n] \to K(\alpha_1, \ldots, \alpha_n)$ be the evaluation homomorphism. Since $g\varphi = \varphi g$ for each $g \in G$ by the proof of Theorem 25.3, $g\varphi(\delta_n) = \varphi(g\delta_n)$. Hence every even element of G leaves $\delta_n(\alpha_1, \ldots, \alpha_n)$ fixed and every odd element of G sends $\delta_n(\alpha_1, \ldots, \alpha_n)$ to its negative.

Now since p has distinct roots, $\varphi(\delta_n) = \delta_n(\alpha_1, \ldots, \alpha_n) \neq 0$. Hence if G contains an odd element, $\delta_n(\alpha_1, \ldots, \alpha_n) \notin \text{Fix}(G) = K$, while if $G \subset A_n$, $\delta_n(\alpha_1, \ldots, \alpha_n) \in \text{Fix}(G) = K$. Since $\pm\delta_n(\alpha_1, \ldots, \alpha_n)$ are the roots of $X^2 - \text{disc}(p)$ in K, we have shown the equivalence of the second item and the third. $\square$

The discriminant resolvent will serve as a prototype for the generality of section 27, where we seek to develop a collection of resolvents to distinguish among the roots of a polynomial and to determine the Galois group of the polynomial. One difficulty in generalizing the preceding proof is that the assumption of distinct roots in the polynomial p was sufficient to conclude that the roots of the discriminant resolvent $X^2 - \text{disc}(p)$ were distinct. As a result, if the resolvent factors, it factors into distinct linear factors, each of multiplicity 1. For more general resolvents, the situation is more complicated, and we will pay particular attention to factorizations of a resolvent which contain a linear factor of multiplicity 1.

26. Exercise Set 1

A.

26.1* Show that the Galois group of a splitting field extension, under the operation of function composition, is a group. Show that $S(N)$, as defined in Definition 24.7, is a group.

26.2* Show that the fixed field of a subgroup $H \subset \text{Gal}(L/K)$ for a splitting field extension L of K is a subfield of L.

26.3* Let K be a subfield of $\mathbb{C}$, L a splitting field over K, and N an intermediate splitting field. Define a map $\pi : \text{Gal}(L/K) \to \text{Gal}(N/K)$ by $\pi(g) = g|_N$, where $g|_N$ denotes the restriction of g to the domain N. Show that π is a homomorphism of groups.

26.4 Let K be a subfield of $\mathbb{C}$ and N a finite normal field extension of K. Prove that N is a splitting field over K without using the fact that N is a simple extension of K. (Hint: First prove that $N = K(\alpha_1, \ldots, \alpha_n)$ for some finite set of $\alpha_i \in N$. Then consider $p = \prod_i m_{\alpha_i, K}$.)

26.5 Determine the group of automorphisms of the splitting field of $X^4 - 2X^2 - 4$ over $\mathbb{Q}$. Prove that this group is isomorphic to the dihedral group of order 8. (Hint: Show that the group is of order 8 and contains an element x of order 4 and an element y of order 2 such that $yxy = x^{-1}$.)

26.6 If T is the orbit of a polynomial $t \in K[X_1, \ldots, X_n]$, explain in terms of cosets how S_n acts on T by permuting the elements of T.

26.7* Let W be a finite set of n elements of a commutative ring R, and set $f(X) = \prod_{w \in W}(X - w) \in R[X]$. Show that the coefficient of X^i in $f(X)$ is $(-1)^{n-i}$ times the sum of all products of $n - i$ distinct elements of W. Show further that this sum is unchanged under any permutation of W.

B.

26.8 Find a field L and proper subfields K_1 and K_2 of L such that the extension L/K_1 is normal but L/K_2 is not normal. (Hint: Work inside a splitting field with Galois group G and subgroups S, T_1, and T_2, such that S is a normal subgroup of T_1 but a nonnormal subgroup of T_2. The splitting field itself, of course, cannot serve as L.)

26.9* Let H be a subgroup of S_n. Prove that $\sum_{h \in H} \prod_{i=1}^{n} X_{h(i)}^i$ has stabilizer H.

26.10* For the ordering of monomials given in the proof of Theorem 25.8, prove that if $f \le f'$ and $g \le g'$, then $fg \le f'g'$.

26.11* Prove a formula for the greatest monomial occurring in the product

$$\sigma_1^{k_1}\sigma_2^{k_2}\cdots\sigma_n^{k_n}$$

of elementary symmetric polynomials in $K[X_1,\ldots,X_n]$.

26.12 Suppose L is a splitting field over a subfield F of $\mathbb{C}$ with Galois group G and K is an intermediate field of L/F, say $K = F(\alpha)$. Let N be the splitting field of $m_{\alpha,F}$, and let $H = \mathrm{Gal}(L/K)$. Prove that $\mathrm{Gal}(L/N) = \cap_{g\in G}\, gHg^{-1}$.

26.13 Suppose L is a splitting field over a subfield F of $\mathbb{C}$ with Galois group G and K is an intermediate field of L/F. Let $H = \mathrm{Gal}(L/K)$ and H' be the normalizer of H in G, that is, the largest subgroup of G in which H is normal. Let $K' = \mathrm{Fix}(H')$. Prove that K/K' is a splitting field. Moreover, prove that if N is an intermediate field of K/F such that K/N is a splitting field, then N contains K'.

26.14* [Natural Irrationalities] Let K be a splitting field over a subfield F of $\mathbb{C}$, and let L be a finite extension of F, say $F(\alpha)$. Set $KL = K(\alpha)$. Prove that KL is a splitting field over L and that $\mathrm{Gal}(KL/L) \cong \mathrm{Gal}(K/K\cap L)$. (Hint: Define a map $\pi : \mathrm{Gal}(KL/L) \to \mathrm{Gal}(K/F)$ by restricting an automorphism of KL to an automorphism of K. Show that this map makes sense and is a homomorphism of groups. Show next that π is one-to-one, so that $\mathrm{Gal}(KL/L)$ is isomorphic to the image of π, which must necessarily take the form $\mathrm{Gal}(K/M)$ for some intermediate field M of K/F. Now show $K\cap L$ is fixed by $\pi(g)$ for every $g \in \mathrm{Gal}(KL/L)$ and hence $K\cap L \subset M$. For the other direction, show that since $\beta \in M$, β is fixed by $\pi(g)$ for every $g \in \mathrm{Gal}(KL/L)$, and hence β is in both K and L and $M \subset K\cap L$.)

27. Distinguishing Numbers, Determining Groups

In section 25, we discovered that the Galois group of a polynomial in $K[X]$ with n distinct roots lies in A_n if and only if the discriminant resolvent factors over K. In this situation, the product δ_n of the differences of the roots lies in K:

$$\delta_n(\alpha_1,\ldots,\alpha_n) = k \tag{27.1}$$

for some $k \in K^*$. This equation expresses a relationship among the roots of p that distinguishes them. More precisely, the same relationship does not hold among the roots if we reorder them by applying an odd permutation s, for sending $\alpha_i \mapsto \alpha_{s(i)}$ negates the left-hand side, but the right-hand side, as an element of the ground field K, must remain fixed. By Proposition 25.1, it is *precisely* the existence of an equation such as (27.1) that determines whether or not the Galois group lies inside A_n. The existence of such equations therefore provides important information in determining the Galois group of the polynomial.

In this section, we ask how the Galois group determines, and is determined by, these sorts of algebraic relations among the roots of a polynomial. In particular, we seek to understand which relations come from polynomials with stabilizers H: given $t \in K[X_1, \ldots, X_n]$ with stabilizer H and a polynomial p of degree n with roots $\alpha_1, \ldots, \alpha_n$, when does $t(\alpha_1, \ldots, \alpha_n) = k$ for some $k \in K$? We will tackle this question by factoring the resolvent $f_{t,p}$ associated to t and p.

At the end of this section, we will use our results to prove the following proposition, which shows how a combination of resolvents and factorization over extension fields – which we had avoided by using resolvents – can efficiently dispose of the determination of the Galois group of a degree 4 polynomial.

Proposition 27.1 ([31]). *Let K be a subfield of $\mathbb{C}$ and*

$$p(X) = X^4 + a_3 X^3 + a_2 X^2 + a_1 X + a_0 \in K[X]$$

an irreducible polynomial of degree 4. Let $G = \mathrm{Gal}(p, K)$, identified with a subgroup of S_n via its action on the roots of p. Let $d(X) = f_{\delta_4, p}(X)$ be the discriminant resolvent and $r(X) = f_{t,p}(X)$ the resolvent for the term $t = X_1 X_2 + X_3 X_4$.

Then G is isomorphic to one of S_4, A_4, D_4, $\mathbb{Z}/4\mathbb{Z}$, or $\mathbb{Z}/2\mathbb{Z} \oplus \mathbb{Z}/2\mathbb{Z}$, as follows:

(1) S_4 if $d(X)$ and $r(X)$ are irreducible over K;

(2) A_4 if $d(X)$ is reducible and $r(X)$ is irreducible over K;

(3) D_4 if $r(X)$ has a unique linear factor $X - b$ over K and the polynomials $X^2 - bX + a_0$ and $X^2 + a_3 X + (a_2 - b)$ do not both factor into linear terms over the splitting field of $r(X)$;

(4) $\mathbb{Z}/4\mathbb{Z}$ if $r(X)$ has a unique linear factor $X - b$ over K and the polynomials $X^2 - bX + a_0$ and $X^2 + a_3 X + (a_2 - b)$ each factor into linear terms over the splitting field of $r(X)$;

(5) $\mathbb{Z}/2\mathbb{Z} \oplus \mathbb{Z}/2\mathbb{Z}$ if $r(X)$ factors into linear terms over K.

27.1. Resolvent Factorization and Conjugacy

First we examine what it means for a stabilizer H of a polynomial t to be a subgroup of the Galois group. As usual, $\varphi \colon K[X_1, \ldots, X_n] \to K(\alpha_1, \ldots, \alpha_n)$ denotes the evaluation map $\varphi(f) = f(\alpha_1, \ldots, \alpha_n)$.

Theorem 27.2 (Resolvent in Fixed Field Theorem I). *Suppose K is a subfield of $\mathbb{C}$, L is a splitting field over K of a polynomial p of degree n with roots $\alpha_1, \ldots, \alpha_n$, and $G = \mathrm{Gal}(L/K)$ is the Galois group, viewed as a subgroup of S_n by virtue of its action on the roots.*

Let $t \in K[X_1, \ldots, X_n]$ be a polynomial with stabilizer $H \subset G \subset S_n$. Then $\varphi(t) = t(\alpha_1, \ldots, \alpha_n) \in \mathrm{Fix}(H)$.

Proof. Since t has stabilizer H, $t(X_1, \ldots, X_n)$ is fixed by every permutation in $H \subset S_n$; moreover, since the action of H on the n "letters" is identified with the action on the α_i, $\varphi(t)$ is fixed by every element of $H \subset G$. By the definition of fixed field, then, $\varphi(t)$ lies in $\mathrm{Fix}(H)$. $\qquad\square$

Next, we consider how linear factors of a resolvent polynomial, which express relationships over K among the roots of the polynomial, aid in identifying when the Galois group lies in a particular class of subgroups.

Definition 27.3 (Conjugacy, Conjugacy Class). Let G be a group. We say that two subgroups H_1 and H_2 of G are *conjugate* if there exists an element $g \in G$ such that $H_1 = g^{-1} H_2 g$. The *conjugacy class* of a subgroup H of G is the set of all subgroups of G which are conjugate to H.

Like *normal*, the word *conjugate* when applied to groups comes from a notion for fields under the Galois correspondence. Let's work the connection out. Suppose that L/K is a splitting field extension for a polynomial $p \in K[X]$ with roots $\alpha_1, \ldots, \alpha_n$, and consider the subgroups $H_1 = \mathrm{Gal}(L/K(\alpha_1))$ and $H_2 = \mathrm{Gal}(L/K(\alpha_2))$ of $\mathrm{Gal}(L/K)$. By the Extension of Isomorphism Theorems (18.6 and 18.7), we know that there exists some automorphism of L over K carrying α_1 to α_2, say $g \in \mathrm{Gal}(L/K)$. (First use $K' = K$, $\tau = id$, $\alpha = \alpha_1$, and $\alpha' = \alpha_2$; then use $L' = L$, $K = K(\alpha_1)$, $K' = K(\alpha_2)$, and τ the isomorphism obtained from the first theorem.) Observe that if $h \in H_2$, then $g^{-1} h g(\alpha_1) = \alpha_1$: g sends α_1 to α_2, h leaves α_2 fixed, and g^{-1} sends α_2 back to α_1. But then $g^{-1} H_2 g \subset H_1$.

Now $[K(\alpha_1) : K] = [K(\alpha_2) : K]$ since both are equal to the degree of the minimal polynomial over K (Theorem 8.6). Then, by the Tower Law (Exercise 13.8), $[L : K(\alpha_1)] =$

$[L : K(\alpha_2)]$. Since the order of a Galois group equals the dimension of the extension (Theorem 24.4), the subgroups H_1 and H_2 have the same order. But then $g^{-1}H_2g$ is of the same order as H_1 and so $g^{-1}H_2g = H_1$. In other words, the Galois groups of L over $K(\alpha_1)$ and $K(\alpha_2)$ – fields generated by conjugate algebraic numbers – are conjugate as subgroups of $\mathrm{Gal}(L/K)$.

It is an exercise (29.1) to show that conjugacy is an equivalence relation. Similarly, it is an exercise (29.9) to show that conjugation in a permutation group on n letters corresponds to renaming the letters.

Theorem 27.4 (Resolvent Factorization Theorem). *Suppose K is a subfield of $\mathbb{C}$, L is a splitting field over K of a polynomial p of degree n with roots $\alpha_1, \ldots, \alpha_n$, and $G = \mathrm{Gal}(L/K)$ is the Galois group, viewed as a subgroup of S_n by virtue of its action on the roots.*

Let $t \in K[X_1, \ldots, X_n]$ be a polynomial with stabilizer $H \subset S_n$. Then

(1) if $s^{-1}Gs \subset H$ for some $s \in S_n$, then $f_{t,p}(X)$ contains a linear factor when factored over K;

(2) if $f_{t,p}(X)$ contains a linear factor of multiplicity 1 when factored over K, then $s^{-1}Gs \subset H$ for some $s \in S_n$. These s are precisely those for which $X - (s(t))(\alpha_1, \ldots, \alpha_n) \in K[X]$ is the linear factor.

Proof. Consider the first item. Suppose that for some $s \in S_n$, $J = s^{-1}Gs \subset H$. Then for each $j \in J$, there exists a $g \in G$ such that $j = s^{-1}gs$, and since $t(X_1, \ldots, X_n)$ is fixed by $j \in J \subset H$,

$$t(X_1, \ldots, X_n) = (j(t))(X_1, \ldots, X_n) = ((s^{-1}gs)(t))(X_1, \ldots, X_n),$$

or $t = (s^{-1}gs)(t) = s^{-1}(gs(t))$. Applying s to both sides, we have $s(t) = gs(t)$. Following by the evaluation map φ, $\varphi(s(t)) = \varphi(gs(t))$, and since $\varphi g = g\varphi$, $\varphi(s(t)) = g\varphi(s(t))$. One checks that as j runs through J, the g in the corresponding elements $s^{-1}gs$ runs through all $g \in G$. Hence the $\varphi(s(t)) = (s(t))(\alpha_1, \ldots, \alpha_n)$ lies in K. Now since $f_{t,p}(X)$ has as the set of its roots the set of $\varphi(s(t)), s \in S_n$, then $f_{t,p}(X)$ has a root in K, $k = \varphi(s(t))$. Therefore we have a linear term $X - k$ appearing in the factorization of $f_{t,p}(X)$ over K (see Exercise 13.1).

For the second item, suppose that $f_{t,p}(X)$ has a linear factor $X - k$ of multiplicity 1 when factored over $K[X]$. Now the roots of $f_{t,p}$ are the $\varphi(u)$ for $u \in T$, where T is the orbit of t. Let u be an element of T such that $\varphi(u) = k$. Observe that $u \in T$ is unique: there is no other $u' \in T$ such that $\varphi(u') = k$, for otherwise the linear factor $X - k$ would have multiplicity greater than 1. In short, $\varphi(u) = \varphi(u')$ implies $u = u'$.

Now let $s \in S_n$ be any element such that $u = s(t)$. By the last paragraph, given any $s' \in S_n$, $\varphi(s'(t)) = k = \varphi(s(t))$ if and only if $s(t) = s'(t)$. But since the elements of the orbit T are in correspondence with the cosets S_n/H, $\varphi(s'(t)) = \varphi(s(t))$ if and only if $sH = s'H$.

Finally, let $g \in G$. Clearly $\varphi(u) = k$ is fixed by g. Then $\varphi(s(t)) = g\varphi(s(t)) = \varphi(gs(t))$. By the foregoing, $sH = gsH$, hence $s^{-1}gs \in H$. Therefore $s^{-1}Gs \subset H$. □

The Resolvent in Fixed Field Theorem provides a method for generating elements of various intermediate fields of L/K, while Resolvent Factorization Theorem provides tools for determining where the Galois group sits inside S_n. In what follows, we use these two theorems to determine subfields and to capture the Galois groups, both without using the factorization of the minimal polynomial over intermediate fields of L/K, as was necessary in Chapter 4.

27.2. Finding Subfields from Linear Factors

Given the Galois group of a splitting field extension L/K of a polynomial $p \in K[X]$, we consider how to determine explicitly the intermediate fields of L/K. By the Galois correspondence, there are exactly as many intermediate fields in L/K as subgroups $H \subset \text{Gal}(L/K)$, and these intermediate fields are the fixed fields in L of these subgroups (Theorem 24.10). What remains, then, is to express these intermediate fields as fields generated over K by some elements, themselves expressed in terms of the roots α_i of p.

To do so, we first prove a more specialized version of Theorem 27.2.

Theorem 27.5 (Resolvent in Fixed Field Theorem II). *Let K be a subfield of $\mathbb{C}$. Suppose p is an irreducible polynomial in $K[X]$ with roots $\alpha_1, \ldots, \alpha_n$, G is the Galois group of the splitting field extension, and $t \in K[X_1, \ldots, X_n]$ is a polynomial with stabilizer H. Let $\beta = t(\alpha_1, \ldots, \alpha_n)$. Then $\beta \in \text{Fix}(H \cap G)$ and $m_{\beta,K}$ divides $f_{t,p}$ in $K[X]$.*

Proof. Now β is clearly fixed by every element of H, so it is fixed by every element of $H \cap G$. Since $H \cap G$ is a subgroup of G, $\text{Fix}(H \cap G)$ is well-defined. The last statement follows from the definition of minimal polynomial, since β is root of $f_{t,p}$. □

In order to find elements in a given intermediate field M of L/K corresponding to a subgroup $N \subset \text{Gal}(L/K)$, we seek polynomials $t \in K[X_1, \ldots, X_n]$ with stabilizer H with the property that Fix($H \cap G$) = M. Then, if $\beta = t(\alpha_1, \ldots, \alpha_n)$, then $K(\beta) \subset M$. There is no guarantee, however, that $K(\beta) = M$, and we may need to find other polynomials t, and hence other field elements β, such that $K(\beta_1, \ldots, \beta_k) = M$. We explore this approach in the following examples. (Note that in section 28, we will see how to use AlgFields to perform the required computations.)

Example 27.6 (Degree 4 Subfield of D_4 Extension). Let $K = \mathbb{Q}$ and consider the polynomial $p(X) = X^4 - 4X^2 - 1 \in K[X]$ with roots

$$\alpha_1 \approx 2.05, \quad \alpha_2 \approx .486i, \quad \alpha_3 \approx -2.05, \quad \text{and} \quad \alpha_4 \approx -.486i.$$

By computation with the Resolvent Factorization Theorem (see section 28 on computation), we will be able to determine that the Galois group G of p corresponds to the subgroup of S_4 of eight elements containing (1234) and (24), isomorphic to the dihedral group with eight elements.

We start by letting H be the cyclic subgroup $\{(), (24)\}$. We seek a description of the subfield corresponding to Fix(H).

Let $t(X_1, \ldots, X_4) = X_1(X_2 + X_4)$. Then one can check that t is a polynomial with stabilizer H and the orbit of t contains twelve terms. Hence the resolvent $f_{t,p}(X)$ is therefore of degree 12. Now we hope that $\beta = t(\alpha_1, \ldots, \alpha_4)$ generates Fix(H) over K, and we search for the minimal polynomial of β among the factors of $f_{t,p}(X)$. However, we find that the minimal polynomial of β is in fact X, since $\beta = 0$.

Instead we let $t(X_1, X_2, X_3, X_4) = X_1$. We check that T has stabilizer W, the subgroup of S_4 consisting of all permutations fixing 1. This subgroup, isomorphic to S_3, has order 6. Using the previous theorem (Theorem 27.5), we find that $\beta = t(\alpha_1, \ldots, \alpha_4) = \alpha_1$ lies in Fix($W \cap G$) = Fix(H), so that even though the t does not have stabilizer H, we may use the β we find anyway. The orbit for t is $T = \{X_1, \ldots, X_4\}$, and the resulting resolvent $f_{t,p}(X)$ is

$$(X - \alpha_1)(X - \alpha_2)(X - \alpha_3)(X - \alpha_4),$$

or $p \in K[X]$! Now $\beta = t(\alpha_1, \ldots, \alpha_4) = \alpha_1$, and $m_{\alpha_1, K}$ should occur as a factor of p. In fact, $m_{\alpha_1, K}$ is p itself. Hence the dimension of $K(\beta)$ over K is 4. Since the dimension of a splitting field extension is the order of the Galois group, we know that [Fix(H) : K] = 4, and hence we have found Fix(H) = $K(\alpha_1)$ (Theorem 24.4).

Example 27.7 (Degree 3 Subfield of S_4 Extension). Let $K = \mathbb{Q}$ and consider the polynomial $p(X) = X^4 + 2X^3 + 3X^2 + 4X + 5 \in K[X]$ with roots

$$\alpha_1 \approx -1.29 - 0.858i, \quad \alpha_2 \approx -1.29 + 0.858i, \quad \alpha_3 \approx 0.288 - 1.42i,$$

$$\text{and } \alpha_4 \approx 0.288 + 1.42i.$$

As before, one may determine that the Galois group G of p is isomorphic to S_4.

Let H be the subgroup of order 8 containing (24), (12)(34), and (14)(23). We seek a generator β over K of the field Fix(H). Since the dimension of a splitting field extension is the order of the Galois group, [Fix(H) : K] = 3 (Theorem 24.4), so we seek a β with minimal polynomial over K of degree 3.

Now we use an idea from Exercise 26.9 to find a polynomial with stabilizer H, choosing the polynomial

$$
\begin{aligned}
t(X_1, \ldots, X_4) = \; & X_1 X_2^2 X_3^3 + X_1 X_3^3 X_4^2 + X_1^2 X_2 X_4^3 + X_2 X_3^2 X_4^3 \\
& + X_2^3 X_3^2 X_4 + X_1^2 X_3^3 X_4 + X_1^3 X_3 X_4^2 + X_1^3 X_2^2 X_3.
\end{aligned}
$$

The corresponding orbit T has three elements, namely t, $g(t)$ where $g = (123) \in S_4$, and $g'(t)$ where $g' = (132) \in S_4$. We form the resolvent

$$f_{t,p}(X) = (X - t(\alpha_1, \ldots, \alpha_4))(X - g(t)(\alpha_1, \ldots, \alpha_4))(X - g'(t)(\alpha_1, \ldots, \alpha_4)),$$

expand and collect coefficients of powers of X, and, expressing these coefficients as symmetric polynomials σ_i in the α_j, we have

$$
\begin{aligned}
f_{t,p}(X) = \; & X^3 + \left(3\sigma_3^2 + 3\sigma_1^2\sigma_4 - 4\sigma_2\sigma_4 - \sigma_1\sigma_2\sigma_3\right)X^2 \\
& + \left(\sigma_2^3\sigma_3^2 + \sigma_1^3\sigma_3^3 - 5\sigma_1\sigma_2\sigma_3^3 + 3\sigma_3^4 + \sigma_1^2\sigma_2^3\sigma_4 - 4\sigma_2^4\sigma_4 - 5\sigma_1^3\sigma_2\sigma_3\sigma_4 \right. \\
& \left. + 16\sigma_1\sigma_2^2\sigma_3\sigma_4 + 2\sigma_1^2\sigma_3^2\sigma_4 - 8\sigma_2\sigma_3^2\sigma_4 + 3\sigma_1^4\sigma_4^2 - 8\sigma_1^2\sigma_2\sigma_4^2\right)X \\
& + \left(-\sigma_1^2\sigma_2^2\sigma_3^4 + 2\sigma_2^3\sigma_3^4 + 2\sigma_1^3\sigma_3^5 - 4\sigma_1\sigma_2\sigma_3^5 + \sigma_3^6 - \sigma_1^4\sigma_2^2\sigma_3^2\sigma_4 \right. \\
& + 8\sigma_1^2\sigma_2^3\sigma_3^2\sigma_4 - 12\sigma_2^4\sigma_3^2\sigma_4 + 2\sigma_1^5\sigma_3^3\sigma_4 - 16\sigma_1^3\sigma_2\sigma_3^3\sigma_4 + 24\sigma_1\sigma_2^2\sigma_3^3\sigma_4 - \sigma_1^2\sigma_3^4\sigma_4 \\
& - 4\sigma_2\sigma_3^4\sigma_4 + 2\sigma_1^4\sigma_2^3\sigma_4^2 - 12\sigma_1^2\sigma_2^4\sigma_4^2 + 16\sigma_2^5\sigma_4^2 - 4\sigma_1^5\sigma_2\sigma_3\sigma_4^2 \\
& \left. + 24\sigma_1^3\sigma_2^2\sigma_3\sigma_4^2 - 32\sigma_1\sigma_2^3\sigma_3\sigma_4^2 - \sigma_1^4\sigma_3^2\sigma_4^2 + 8\sigma_1^2\sigma_2\sigma_3^2\sigma_4^2 + \sigma_1^6\sigma_4^3 - 4\sigma_1^4\sigma_2\sigma_4^3\right).
\end{aligned}
$$

Now, writing the polynomial $p(X) = X^4 + \sum_{i=0}^{3} c_i X^i$ and substituting $(-1)^i c_{n-i}$ for σ_i as in equation (25.4), we deduce that $f_{t,p}(X) = X^3 + 24X^2 - 408X - 5088$. Since this polynomial is irreducible, $[K(\beta) : K] = 3$. Then, since $[L : \text{Fix}(H)] = |H| = 8$ and by the Tower Law, $[\text{Fix}(H) : K] = 3$, we conclude that $K(\beta) = \text{Fix}(H)$.

We calculate that $\beta \approx -9.34$. Note that the other roots of the same polynomial ($\approx -31.80, 17.14$) will generate $\text{Fix}(s^{-1}Hs)$ for the other conjugates $s^{-1}Hs$ of H in S_4. Why?

27.3. Finding the Galois Group from Linear Factors

Now we turn to the determination of the Galois group with Galois resolvents. First we show that the Galois group of an irreducible polynomial $p \in K[X]$ sits inside the corresponding symmetric group in a special way.

Definition 27.8 (Transitive Subgroup of S_n). A subgroup H of S_n is *transitive*, if given any two elements $i, j \in \{1, 2, \ldots, n\}$, there exists some permutation $h \in H$ such that $h(i) = j$.

Proposition 27.9. *Let K be a subfield of $\mathbb{C}$. Let $p \in K[X]$ be a polynomial of degree n that is irreducible over K with roots $\alpha_1, \ldots, \alpha_n$, and L the splitting field of p over K. Then, identified with the corresponding permutations in S_n, $\text{Gal}(L/K)$ is a transitive subgroup.*

Proof. Let i, j be as in the definition of transitive. By the Extension of Isomorphism Theorem (18.6), $K(\alpha_i) \cong K(\alpha_j)$ under an isomorphism g extending the identity on K and such that $g(\alpha_i) = \alpha_j$. By the Extension of Isomorphism to Splitting Field Theorem (18.7), there is an automorphism $\tilde{g} \in \text{Gal}(L/K)$ extending g. Hence, under the identification of $\text{Gal}(L/K)$ with a subgroup of S_n where roots α_i are associated to integers i, $\tilde{g}$ corresponds to a permutation sending i to j. $\square$

To determine the Galois group of an irreducible polynomial $p \in K[X]$ of degree n, we may use the second item of the Resolvent Factorization Theorem (27.4) as follows:

(1) Determine all transitive subgroups of S_n.
(2) For each transitive subgroup H, find a corresponding polynomial t with stabilizer H and construct $f_{t,p}(X)$ for each.
(3) Factor each $f_{t,p}$ over K. By the Resolvent Factorization Theorem (27.4), the Galois group $\text{Gal}(L/K)$ is contained in some conjugate of H if $f_{t,p}$ contains a linear factor of multiplicity 1. If necessary, determine the exact conjugate of H by detecting by approximation which element(s) of the orbit of t give linear factors in $f_{t,p}$.

If $p \in K[X]$ is not irreducible, one alters the algorithm to consider intransitive subgroups of S_n.

This procedure is, of course, impossible to execute by hand, so in the next sections, we introduce some computational tools together with examples of using this procedure. We can, however, use the result in proving Proposition 27.1, stated at the beginning of the section.

Proof of Proposition 27.1. Let L denote the splitting field of p over K and $\alpha_1, \ldots, \alpha_4$ the roots of p. We identify G with the subgroup of S_4 given by its action on the roots $\alpha_1, \ldots, \alpha_4$.

We first consider $r(X) = f_{t,p}(X)$ for $t = X_1 X_2 + X_3 X_4$. Referring to the list of subgroups of S_4 in the appendix, observe that the stabilizer of the polynomial t is the subgroup of S_4 generated by the permutations (1324) and (12). Denote this subgroup, which is isomorphic to D_4, by $D_{4,1}$.

Next we show that $r(X)$ has distinct roots. It is straightforward to check that $\mathrm{disc}(r) = \mathrm{disc}(p)$, by factoring the differences of the elements in the orbit of t. Since p is an irreducible polynomial, p has four distinct roots, and hence $\mathrm{disc}(p) \neq 0$. Then $\mathrm{disc}(r) \neq 0$, and hence $r(X)$ has three distinct roots.

Now we consider possibilities for G. Because $p \in K[X]$ is irreducible, G is transitive (Proposition 27.9). Surveying the transitive subgroups of S_4, we find that G must be isomorphic to one of the five possibilities

$$S_4, \quad A_4, \quad D_4, \quad \mathbb{Z}/4\mathbb{Z}, \quad \text{and} \quad \mathbb{Z}/2\mathbb{Z} \oplus \mathbb{Z}/2\mathbb{Z}. \tag{27.2}$$

We show that $r(X)$ is irreducible over K if and only if $G \cong A_4$ or $G \cong S_4$. Suppose first that $r(X)$ is irreducible over K. Let β be a root of r. Then $[K(\beta) : K] = 3$ and by the Tower Law (Exercise 13.8), $[L : K]$ is divisible by 3. Hence $G \cong A_4$ or $G \cong S_4$. Going the other way, assume $G \cong A_4$ or $G \cong S_4$. If $r(X)$ is reducible, then it must have a linear factor, and, since all of the roots are distinct, any linear factor has multiplicity 1. Hence, by the Resolvent Factorization Theorem (27.4), $s^{-1}Gs \subset D_{4,1}$ for some $s \in S_n$. But then G has at most eight elements, a contradiction. Hence $r(X)$ is irreducible over K.

If $G \cong A_4$ or $G \cong S_4$, the reducibility or irreducibility over K of the discriminant resolvent $d(X)$ determines G precisely, by Proposition 25.1.

Now suppose that $r(X) = f_{t,p}(X)$ is reducible over K. We have already established that $s^{-1}Gs \subset D_{4,1}$ for some $s \in S_n$, or, equivalently, $G \subset s D_{4,1} s^{-1}$ for some $s \in S_n$.

By a variant of the Resolvent Factorization Theorem in Exercise 29.10, if $G \subset s D_{4,1} s^{-1}$ for all $s \in S_4$, then $r(X)$ factors into linear terms over K, and if $r(X)$ factors into distinct linear terms over K, then $G \subset sD_{4,1}s^{-1}$ for every $s \in S_4$. We calculate the intersection of all the subgroups in the conjugacy class of $D_{4,1}$ to be $V = \{(), (12)(34), (13)(24), (14)(23)\} \cong \mathbb{Z}/2\mathbb{Z} \oplus \mathbb{Z}/2\mathbb{Z}$. But if G lies in this intersection, then by checking orders of elements in the the list (27.2) above, we have $G = V$. Again, since we know $r(X)$ has distinct roots, we conclude that $G = V$ (and hence $G \cong \mathbb{Z}/2\mathbb{Z} \oplus \mathbb{Z}/2\mathbb{Z}$) if and only if $r(X)$ factors into linear terms.

Now assume that $r(X)$ has precisely one linear factor over K: $X - b$. Then $b = (s(t))(\alpha_1, \ldots, \alpha_4)$ for some $s \in S_4$; assume for now that s is the identity, for the other cases follow similarly. We are left with the two possibilities $G \cong D_4$ and $G \cong \mathbb{Z}/4\mathbb{Z}$. Let K' be the splitting field of $r(X)$ over K. We deduce that $[K' : K] = 2$; hence by the Tower Law $[L : K']$ is 4 or 2, corresponding to the two possibilities for the isomorphism class of G.

Using Exercise 26.7, one shows that the roots of $X^2 - bX + a_0$ are $\alpha_1\alpha_2$ and $\alpha_3\alpha_4$ and that the roots of $X^2 + a_3X + (a_2 - b)$ are $\alpha_1 + \alpha_2$ and $\alpha_3 + \alpha_4$.

If both polynomials split into linear factors over K', then each of these roots lies in K'. Moreover, we check that $(X - \alpha_1)(X - \alpha_2) \in K'[X]$. Let $M = K'(\alpha_1, \alpha_2)$, and then $[M : K'] \le 2$. Now the two roots $\alpha_1\alpha_3 + \alpha_2\alpha_4$ and $\alpha_1\alpha_4 + \alpha_2\alpha_3$ of $r(X)$ are contained in K' and hence the difference $(\alpha_1 - \alpha_2)(\alpha_3 - \alpha_4) \in K'$ as well. But then $\alpha_3 - \alpha_4 \in M$. Since $\alpha_3 - \alpha_4$ and $\alpha_3 + \alpha_4 \in M$, we deduce that $\alpha_3, \alpha_4 \in M$. But then all the roots of p lie in M and so $[L : K'] = 2$, whence $G \cong \mathbb{Z}/4\mathbb{Z}$.

Going the other way, suppose that $G \cong \mathbb{Z}/4\mathbb{Z}$. Then the quadratic extension K'/K is the only quadratic extension of K lying in L. If one of the polynomials $X^2 - bX + a_0$ and $X^2 + a_3X + (a_2 - b)$ is not split over K, the corresponding splitting field is a quadratic extension of K inside L, hence K'. But then each must split into linear factors over K'. Hence if $G \cong \mathbb{Z}/4\mathbb{Z}$ then the two polynomials split into linear factors over the splitting field of $r(X)$, and we are done. $\qquad\square$

28. Computation of Galois Groups and Resolvents: *Maple* and *Mathematica*

28.1. The Galois Group, I

In the last chapter we introduced enough tools in `AlgFields` to approach the problem of determining the Galois group of a splitting field extension. In fact, `AlgFields` contains a function `FGaloisGroup` that uses the methods of the last chapter to determine the Galois group. (It is important that the field given as the argument have been declared as a splitting field, and if two fields are given as arguments, the first must have been declared as a splitting field extension of the second.) The function returns a list of pairs, where each pair represents an automorphism: each pair is a pair of lists, where under the automorphism each root named in the first list is mapped to the root in the corresponding position in the second list.

We use the function to calculate the Galois groups of the first two cubic examples (17.1 and 17.2):

```
>  FDeclareSplittingField(C1,
>     X^3+3*X+1,[alpha1,alpha2,alpha3]);
----Details of field C1 ----
  Algebraic Numbers: [alpha1, alpha2,
    alpha3]

  Dimension over Q: 6

  Minimal Polynomials: [
    alpha1^3+3*alpha1+1,
    alpha2^2+alpha1*alpha2+3+alpha1^2,
    alpha3+alpha2+alpha1]

  Root Approximations: [

    .16109267731304279646+

      1.7543809597837216610*I,

    -.32218535462608559290-

      .16852902778432639141e-20*I,

    .16109267731304279644-

      1.7543809597837216610*I]

----
```

$In[1] :=$ FDeclareSplittingField[C1,

$$X\hat{\ }3 + 3X + 1, \{\alpha 1, \alpha 2, \alpha 3\}]$$

----Details of field C1 ----

Algebraic Numbers: $\{\alpha 1, \alpha 2, \alpha 3\}$

Dimension over Q: 6

Minimal Polynomials: {

$$1 + 3 * \alpha 1 + \alpha 1\hat{\ }3,$$

$$3 + \alpha 1\hat{\ }2 + \alpha 1 * \alpha 2 + \alpha 2\hat{\ }2,$$

$$\alpha 1 + \alpha 2 + \alpha 3\}$$

Root Approximations: {

$$-0.32218535462608559291,$$

$$0.16109267731304279646 -$$

$$1.7543809597837216610 \ I,$$

$$0.16109267731304279646 +$$

$$1.7543809597837216610 \ I\}$$

```
>  FGaloisGroup(C1);

    [[[α1, α2, α3], [α1, α2, α3]],
     [[α1, α2, α3], [α1, α3, α2]],
     [[α1, α2, α3], [α2, α1, α3]],
     [[α1, α2, α3], [α2, α3, α1]],
     [[α1, α2, α3], [α3, α1, α2]],
     [[α1, α2, α3], [α3, α2, α1]]]
```

$In[2] :=$ FGaloisGroup[C1]

$Out[2]=$ {{$\{\alpha 1, \alpha 2, \alpha 3\}, \{\alpha 1, \alpha 2, \alpha 3\}$},

$\{\{\alpha 1, \alpha 2, \alpha 3\}, \{\alpha 1, \alpha 3, \alpha 2\}\},$

$\{\{\alpha 1, \alpha 2, \alpha 3\}, \{\alpha 2, \alpha 1, \alpha 3\}\},$

$\{\{\alpha 1, \alpha 2, \alpha 3\}, \{\alpha 2, \alpha 3, \alpha 1\}\},$

$\{\{\alpha 1, \alpha 2, \alpha 3\}, \{\alpha 3, \alpha 1, \alpha 2\}\},$

$\{\{\alpha 1, \alpha 2, \alpha 3\}, \{\alpha 3, \alpha 2, \alpha 1\}\}\}$

```
>  FDeclareSplittingField(C2,
>     X^3-3*X-1,[beta1,beta2,beta3]);

----Details of field C2 ----

  Algebraic Numbers: [beta1, beta2, beta3]

  Dimension over Q: 3

  Minimal Polynomials: [
     beta1^3-3*beta1-1,
     beta2+2+beta1-beta1^2,
     beta3-2+beta1^2]

  Root Approximations: [
     1.8793852415718167681,
     -.3472963553338606977,
     -1.5320888862379560704]

----
```

```
In[3]:= FDeclareSplittingField[C2,
           X^3 - 3X - 1, {β1, β2, β3}]

----Details of field C2 ----

  Algebraic Numbers: {β1, β2, β3}

  Dimension over Q: 3

  Minimal Polynomials: {
     -1 - 3 * β1 + β1^3,
     -2 + β1^2 + β2,
     2 + β1 - β1^2 + β3}

  Root Approximations:  {
     -1.5320888862379560704,
     -0.34729635533386069770,
     1.8793852415718167681}

----
```

```
>  FGaloisGroup(C2);

     [[[β1, β2, β3], [β1, β2, β3]],
      [[β1, β2, β3], [β2, β3, β1]],
      [[β1, β2, β3], [β3, β1, β2]]]
```

```
In[4]:= FGaloisGroup[C2]

Out[4]= {{{β1, β2, β3}, {β1, β2, β3}},
         {{β1, β2, β3}, {β2, β3, β1}},
         {{β1, β2, β3}, {β3, β1, β2}}}
```

Sometimes it is of interest to determine the action of the Galois group on algebraic numbers that have different minimal polynomials. In this case, we may declare a splitting field using the polynomial that is the product of the various minimal polynomials – hence a *reducible* polynomial. We may then execute FGaloisGroup to determine the automorphisms of the splitting field expressed as permutations of all of the roots of the reducible polynomial. As an example, we use the second cubic example above and additionally determine the action of the Galois group on the square root of the discriminant.

```
>  FMinPoly((beta1-beta2)*(beta1-beta3)*
>     (beta2-beta3),X,C1);
           135 + X^2
```

```
In[5]:= FMinPoly[(β1 - β2) * (β1 - β3)*
           (β2 - β3), X, C1]
Out[5]= 135 + X^2
```

```
>   FDeclareSplittingField(C1,
>       (X^3+3*X+1)*(X^2+135),
>       [alpha1,delta1,alpha2,delta2,alpha3]);
----Details of field C1 ----
    Algebraic Numbers: [alpha1, delta1, alpha2,
        delta2, alpha3]
    Dimension over Q: 6
    Minimal Polynomials: [
        alpha1^3+3*alpha1+1,
        135+delta1^2,
        alpha2+(2/15-1/30*alpha1+
            1/15*alpha1^2)*delta1+1/2*alpha1,
        delta2+delta1,
        alpha3+(-2/15+1/30*alpha1-
            1/15*alpha1^2)*delta1+1/2*alpha1]
    Root Approximations: [
        .16109267731304279646+
            1.7543809597837216610*I,
        11.618950038622250656*I,
        -.32218535462608559292+
            .166e-18*I,
        -11.618950038622250656*I,
        .16109267731304279646-
            1.7543809597837216612*I]
----
```

$In[6] :=$ FDeclareSplittingField[C1,
$$(X^3 + 3X + 1) * (X^2 + 135),$$
$$\{\alpha1, \gamma1, \gamma2, \alpha2, \alpha3\}]$$

```
----Details of field C1 ----
Algebraic Numbers: {α1, γ1, γ2, α2, α3}
Dimension over Q: 6
Minimal Polynomials: {
    1 + 3 * α1 + α1^3,
    135 + γ1^2,
    γ1 + γ2,
    α1/2 + (-2/15 + α1/30 -
        α1^2/15) * γ1 + α2,
    α1/2 + (2/15 - α1/30 +
        α1^2/15) * γ1 + α3}
Root Approximations: {
    -0.32218535462608559291,
    -11.618950038622250656 I,
    11.618950038622250656 I,
    0.16109267731304279646 -
        1.7543809597837216610 I,
    0.16109267731304279646 +
        1.7543809597837216610 I}
----
```

```
>   FGaloisGroup(C1);
        [[[α1, δ1, α2, δ2, α3], [α1, δ1, α2, δ2, α3]],
        [[α1, δ1, α2, δ2, α3], [α1, δ2, α3, δ1, α2]],
        [[α1, δ1, α2, δ2, α3], [α2, δ1, α3, δ2, α1]],
        [[α1, δ1, α2, δ2, α3], [α2, δ2, α1, δ1, α3]],
        [[α1, δ1, α2, δ2, α3], [α3, δ1, α1, δ2, α2]],
        [[α1, δ1, α2, δ2, α3], [α3, δ2, α2, δ1, α1]]]]
```

$In[7] :=$ FGaloisGroup[C1]
$Out[7] =$ {{{α1, γ1, γ2, α2, α3}, {α1, γ1, γ2, α2, α3}},
{{α1, γ1, γ2, α2, α3}, {α1, γ2, γ1, α3, α2}},
{{α1, γ1, γ2, α2, α3}, {α2, γ1, γ2, α3, α1}},
{{α1, γ1, γ2, α2, α3}, {α2, γ2, γ1, α1, α3}},
{{α1, γ1, γ2, α2, α3}, {α3, γ1, γ2, α1, α2}},
{{α1, γ1, γ2, α2, α3}, {α3, γ2, γ1, α2, α1}}}

Once the degree of the field extension becomes larger, however, it can be difficult to factor polynomials over subfield extensions, and then the methods of the last chapter become inefficient. For instance, even when the degree of an irreducible polynomial reaches 5, FGaloisGroup may take quite a long time to determine the Galois group. In this case, we may turn to the methods of this chapter.

28.2. Galois Resolvents

Once we have chosen a polynomial with stabilizer H for some subgroup H of S_n, say by taking a product of powers of $X_1, \ldots, X_n$ and summing over the images under the permutations in H, then AlgFields can determine the orbit and the resolvent itself.

The function FPolynomialOrbit takes a polynomial and a list of variables, returning the set of all permutations of the variables, applied to the polynomial. If a third argument is present, then the variables in the list are substituted with numerical approximations to the roots of the third argument, taken in root number order.

`> FPolynomialOrbit(X1*X2^2,[X1,X2,X3]);`	*In[8]:=* FPolynomialOrbit[X1 * X2^2, {X1, X2, X3}]
$[X1\ X2^2,\ X2\ X1^2,\ X3\ X1^2,\ X1\ X3^2,$ $X2\ X3^2,\ X3\ X2^2]$	*Out[8]=* {X1 X2^2, X1 X3^2, X1^2 X2, X2 X3^2, X1^2 X3, X2^2 X3}

`> FPolynomialOrbit(X1*X2^2,[X1,X2,X3],` `>    X^3-3*X-1);`	*In[9]:=* FPolynomialOrbit[X1 * X2^2, {X1, X2, X3}, X^3 − 3X − 1]
$[0.2266815969, -1.226681597,$ $-5.411474130,\ 4.411474128,$ $-0.8152074689, -0.1847925308]$	*Out[9]=* {−0.184793, −5.41147, −0.815207, −1.22668, 4.41147, 0.226682}

The function FGaloisResolvent takes the same information together with a variable name and returns the resolvent with coefficients expressed in the elementary symmetric functions $\sigma_1, \ldots, \sigma_n$. If the fourth argument is the minimal polynomial of the roots α_i, the function will substitute the correct values of the symmetric functions using the coefficients of the polynomial.

```
>   FGaloisResolvent(X1*X2^2,[X1,X2,X3],
>       X);
```

$$\sigma 3^6 + X(-\sigma 1\,\sigma 2\,\sigma 3^4 + 3\,\sigma 3^5)+$$
$$X^2\,(\sigma 1^3\,\sigma 3^3 + \sigma 2^3\,\sigma 3^2 - 5\,\sigma 1\,\sigma 2\,\sigma 3^3 + 6\,\sigma 3^4)+$$
$$X^3\,(-\sigma 1^2\,\sigma 2^2\,\sigma 3 + 2\,\sigma 1^3\,\sigma 3^2+$$
$$2\,\sigma 2^3\,\sigma 3 - 6\,\sigma 1\,\sigma 2\,\sigma 3^2 + 7\,\sigma 3^3)+$$
$$X^4\,(\sigma 1^3\,\sigma 3 + \sigma 2^3 - 5\,\sigma 1\,\sigma 2\,\sigma 3 + 6$$
$$\sigma 3^2)+$$
$$X^5\,(-\sigma 1\,\sigma 2 + 3\,\sigma 3) + X^6$$

```
In[10]:= FGaloisResolvent[X1 * X2^2, {X1, X2, X3},
         X]
```

$$Out[10]= X^6 + \sigma 3^6 + X^5\,(-\sigma 1\,\sigma 2 + 3\,\sigma 3)+$$
$$X^4\,(\sigma 2^3 + \sigma 1^3\,\sigma 3 - 5\,\sigma 1\,\sigma 2\,\sigma 3 + 6\,\sigma 3^2)+$$
$$X^3\,(-\sigma 1^2\,\sigma 2^2\,\sigma 3 + 2\,\sigma 2^3\,\sigma 3+$$
$$2\,\sigma 1^3\,\sigma 3^2 - 6\,\sigma 1\,\sigma 2\,\sigma 3^2 + 7\,\sigma 3^3)+$$
$$X^2\,(\sigma 2^3\,\sigma 3^2 + \sigma 1^3\,\sigma 3^3 - 5\,\sigma 1\,\sigma 2\,\sigma 3^3+$$
$$6\,\sigma 3^4)+$$
$$X\,(-\sigma 1\,\sigma 2\,\sigma 3^4 + 3\,\sigma 3^5)$$

```
>   FGaloisResolvent(X1*X2^2,[X1,X2,X3],
>       X,X^3-3*X-1);
```

$$1 + 3\,X - 21\,X^2 - 47\,X^3-$$
$$21\,X^4 + 3\,X^5 + X^6$$

```
>   factor(%);
```

$$(X^3 - 3\,X^2 - 6\,X - 1)$$
$$(X^3 + 6\,X^2 + 3\,X - 1)$$

```
In[11]:= FGaloisResolvent[X1 * X2^2, {X1, X2, X3},
         X, X^3 - 3X - 1]
```

$$Out[11]= 1 + 3\,X - 21\,X^2 - 47\,X^3-$$
$$21\,X^4 + 3\,X^5 + X^6$$

```
In[12]:= Factor[%]
```

$$Out[12]= (-1 - 6\,X - 3\,X^2 + X^3)$$
$$(-1 + 3\,X + 6\,X^2 + X^3)$$

```
>   FGaloisResolvent(X1*X2^2,[X1,X2,X3],
>       X,X^3+3*X+1);
```

$$1 - 3\,X + 33\,X^2 - 61\,X^3+$$
$$33\,X^4 - 3\,X^5 + X^6$$

```
>   factor(%);
```

$$1 - 3\,X + 33\,X^2 - 61\,X^3+$$
$$33\,X^4 - 3\,X^5 + X^6$$

```
In[13]:= FGaloisResolvent[X1 * X2^2, {X1, X2, X3},
         X, X^3 + 3X + 1]
```

$$Out[13]= 1 - 3\,X + 33\,X^2 - 61\,X^3+$$
$$33\,X^4 - 3\,X^5 + X^6$$

```
In[14]:= Factor[%]
```

$$Out[14]= 1 - 3\,X + 33\,X^2 - 61\,X^3+$$
$$33\,X^4 - 3\,X^5 + X^6$$

```
>    FGaloisResolvent((X1 - X2)*(X1 - X3)*
>        (X1 - X4)*(X2 - X3)*(X2 - X4)*
>        (X3 - X4),[X1, X2, X3, X4], X,
>        X^4 - 10*X^2 - 122);
```

$$674892288 + X^2$$

```
>    factor(%);
```

$$674892288 + X^2$$

```
In[15]:= FGaloisResolvent[(X1 − X2) * (X1 − X3)*
             (X1 − X4) * (X2 − X3) * (X2 − X4)*
             (X3 − X4), {X1, X2, X3, X4}, X,
             X^4 − 10X^2 − 122]

Out[15]= 674892288 + X^2

In[16]:= Factor[%]

Out[16]= 674892288 + X^2
```

Since nearly all of the work is in reducing the coefficients of a resolvent to polynomials in the symmetric functions in those variables, we may want to calculate a resolvent once and use it for many polynomials. In this case, if the resolvent will be very large, we should store the result of FGaloisResolvent, using, so as not to waste time printing the result, either a colon (*Maple*) or a semicolon (*Mathematica*) at the end of the statement. Then, we may use FSubstituteInGaloisResolvent to perform the substitutions of the symmetric functions given coefficients of particular polynomials.

```
>    myc4resolv := FGaloisResolvent(
>        X1*(X2 + X2^2)*(2*X3 + X3^2) +
>        (X1 + X1^2)*(2*X2 + X2^2)*X4 +
>        (2*X1 + X1^2)*X3*(X4 + X4^2) +
>        X2*(X3 + X3^2)*(2*X4 + X4^2),
>        [X1, X2, X3, X4], X ):

>    factor(
>        FSubstituteInGaloisResolvent(
>        myc4resolv, X^4 - 10*X^2 - 122));
```

$$(X^4 − 976\,X^3 + 2011536\,X^2 +$$
$$1235014784\,X + 2469934548544)$$
$$(X^2 − 1952\,X + 1239520)$$

```
In[17]:= myc4resolv = FGaloisResolvent[
             X1 (X2 + X2^2) (2 X3 + X3^2)+
             (X1 + X1^2) (2 X2 + X2^2) X4+
             (2 X1 + X1^2) X3 (X4 + X4^2)+
             X2 (X3 + X3^2) (2 X4 + X4^2),
             {X1, X2, X3, X4}, X];


         Factor[
            FSubstituteInGaloisResolvent[
             myc4resolv, X^4 − 10X^2 − 122]]

Out[17]= (1239520 − 1952 X + X^2)
            (2469934548544 + 1235014784 X+
             2011536 X^2 − 976 X^3 + X^4)
```

28.3. The Galois Group, II

Using these tools, we may illustrate how one may find the Galois group of a polynomial using the technique of the previous section – that is, without factoring polynomials over algebraic number fields.

Now for the example below, it is actually faster to execute FGaloisGroup. However, almost all of the time spent in executing the method below is due to the one-time calculation of the resolvent formula – the resolvent with coefficients reduced into polynomials in the symmetric functions. Once calculated, though, these resolvent formulas can be used over and over again to determine whether other polynomials of the same degree have Galois groups lying inside certain subgroups.

Let $K = \mathbb{Q}$, $p(X) = X^4 - 10X^2 - 122 \in K[X]$, and L be the splitting field of p over K. Then $\mathrm{Gal}(L/K)$ is a transitive subgroup of S_4, with the particular identification dependent on the numbering of the four roots.

From the appendix, we see that there are five isomorphism classes of transitive subgroups of S_4. The first class, all isomorphic to the cyclic group on four letters, contains three representatives: $C_{4,1} = \langle(1234)\rangle$; $C_{4,2} = \langle(1342)\rangle$; and $C_{4,3} = \langle(1423)\rangle$. The second, all isomorphic to $(\mathbb{Z}/2\mathbb{Z})^2$, contains one representative: $Z_2^2 = \langle(13)(24), (12)(34)\rangle$. (Note: there is another conjugacy class of subgroups isomorphic to $(\mathbb{Z}/2\mathbb{Z})^2$, but these are not transitive.) The third, all isomorphic to the dihedral group of order 8, contains three representatives: $D_{4,1} = \langle(12), (34), (13)(24)\rangle$; $D_{4,2} = \langle(13), (24), (14)(23)\rangle$; and $D_{4,3} = \langle(14), (23), (12)(34)\rangle$. The fourth class contains the subgroups isomorphic to A_4, and there is but one subgroup of S_4 isomorphic to A_4, namely A_4 itself. The last class consists of S_4 itself.

Our first goal is to determine if $\mathrm{Gal}(L/K) \subset A_4$. Our polynomial with stabilizer A_4 is

$$t(X_1, \ldots, X_4) = (X_1 - X_2)(X_1 - X_3)(X_1 - X_4)(X_2 - X_4)(X_3 - X_4),$$

and the corresponding resolvent, after expressing the coefficients as symmetric polynomials in the α_i and substituting $\sigma_1 = 0$, $\sigma_2 = -10$, $\sigma_3 = 0$, $\sigma_4 = -122$, is $X^2 + 674892288$, which is irreducible:

```
> FGaloisResolvent((X1 - X2)*(X1 - X3)*
>     (X1 - X4)*(X2 - X3)*(X2 - X4)*
>     (X3 - X4),[X1, X2, X3, X4],
>     X, X^4 - 10*X^2 - 122);
```
$$674892288 + X^2$$
```
> factor(%);
```
$$674892288 + X^2$$

```
In[18]:= FGaloisResolvent[(X1 − X2) * (X1 − X3)*
         (X1 − X4) * (X2 − X3) * (X2 − X4)*
         (X3 − X4), {X1, X2, X3, X4},
         X, X^4 − 10X^2 − 122]
```
$$Out[18]= 674892288 + X^2$$
```
In[19]:= Factor[%]
```
$$Out[19]= 674892288 + X^2$$

Hence $\mathrm{Gal}(L/K) \not\subset A_4$. As a result, $\mathrm{Gal}(L/K)$ is not isomorphic to A_4 and is either cyclic of order 4, noncyclic abelian of order 4, or dihedral of order 8.

We next decide if $\mathrm{Gal}(L/K) \subset C_{4,i}$ for some i. We choose the polynomial $X_1 X_2^2 X_3^3 + X_2 X_3^2 X_4^3 + X_1^3 X_3 X_4^2 + X_1^2 X_2^3 X_4$ with stabilizer $C_{4,1}$, by taking $X_1 X_2^2 X_3^2$ and summing over all of the images of this polynomial under the permutations in $C_{4,1}$. The Galois resolvent has degree 6 and may be calculated as follows:

```
> factor(FGaloisResolvent(
>    X1*X2^2*X3^3 + X2*X3^2*X4^3 +
>    X1^3*X3*X4^2 + X1^2*X2^3*X4,
>    [X1, X2, X3, X4], X,
>    X^4 - 10*X^2 - 122));
```

$$(X - 2440)^2 (X^2 + 1220000)^2$$

```
In[20]:= Factor[FGaloisResolvent[
    X1 * X2^2 * X3^3 + X2 * X3^2 * X4^3+
    X1^3 * X3 * X4^2 + X1^2 * X2^3 * X4,
    {X1, X2, X3, X4}, X,
    X^4 - 10X^2 - 122]]
Out[20]= (-2440 + X)^2 (1220000 + X^2)^2
```

Unfortunately we encounter a linear factor with multiplicity 2, so we cannot use the Resolvent Factorization Theorem (27.4) to determine whether or not $\mathrm{Gal}(L/K) \subset C_{4,i}$ for some i.

Instead we choose the polynomial

$$t(X_1, X_2, X_3, X_4) = X_1(X_2 + X_2^2)(2X_3 + X_3^2) + (X_1 + X_1^2)(2X_2 + X_2^2)X_4$$
$$+ (2X_1 + X_1^2)X_3(X_4 + X_4^2) + X_2(X_3 + X_3^2)(2X_4 + X_4^2)$$

with stabilizer $C_{4,1}$. The resolvent has degree 6 and may be calculated and factored as follows:

```
> factor(FGaloisResolvent(
>    X1*(X2 + X2^2)*(2*X3 + X3^2) +
>    (X1 + X1^2)*(2*X2 + X2^2)*X4 +
>    (2*X1 + X1^2)*X3*(X4 + X4^2) +
>    X2*(X3 + X3^2)*(2*X4 + X4^2),
>    [X1, X2, X3, X4], X,
>    X^4 - 10*X^2 - 122));
```

$$(X^4 - 976\,X^3 + 2011536\,X^2 + 1235014784\,X +$$
$$2469934548544)$$
$$(X^2 - 1952\,X + 1239520)$$

```
In[21]:= Factor[FGaloisResolvent[
    X1 (X2 + X2^2) (2 X3 + X3^2)+
    (X1 + X1^2) (2 X2 + X2^2) X4+
    (2 X1 + X1^2) X3 (X4 + X4^2)+
    X2 (X3 + X3^2) (2 X4 + X4^2),
    {X1, X2, X3, X4}, X,
    X^4 - 10X^2 - 122]]
Out[21]= (1239520 - 1952 X + X^2)
    (2469934548544 + 1235014784 X+
    2011536 X^2 - 976 X^3 + X^4)
```

Since the factorization over K has no linear factor, the Galois group $\mathrm{Gal}(L/K)$ is not contained in any of the subgroups $C_{4,i}$. Considering the inclusion relations among the transitive subgroups of S_4, we find that the Galois group is one of the subgroups $D_{4,i}$.

In order to determine which one, we consider a resolvent again, this time from the polynomial

$$t(X_1, X_2, X_3, X_4) = X_1^3 X_2^2 X_3 + X_1 X_2^2 X_3^3 + X_1^2 X_2^3 X_4 + X_2^3 X_3^2 X_4$$
$$+ X_1^3 X_3 X_4^2 + X_1 X_3^3 X_4^2 + X_1^2 X_2 X_4^3 + X_2 X_3^2 X_4^3;$$

with stabilizer $D_{4,2}$. As before, we calculate the factorization of the resolvent $f_{t,p}(X)$:

```
> factor(FGaloisResolvent(
>     X1*X2^2*X3^3 + X2*X3^2*X4^3 +
>     X3*X4^2*X1^3 + X4*X1^2*X2^3 +
>     X3*X2^2*X1^3 + X2*X1^2*X4^3 +
>     X1*X4^2*X3^3 + X4*X3^2*X2^3,
>     [X1, X2, X3, X4], X,
>     X^4 - 10*X^2 - 122));
```

$$(X - 4880)(X^2 + 4880000)$$

```
In[22]:= Factor[FGaloisResolvent[
    X1 * X2^2 * X3^3 + X2 * X3^2 * X4^3+
    X3 * X4^2 * X1^3 + X4 * X1^2 * X2^3+
    X3 * X2^2 * X1^3 + X2 * X1^2 * X4^3+
    X1 * X4^2 * X3^3 + X4 * X3^2 * X2^3,
    {X1, X2, X3, X4}, X,
    X^4 - 10X^2 - 122]]
```

$$Out[22] = (-4880 + X)(4880000 + X^2)$$

In order to determine which of the three groups $D_{4,i}$ is the Galois group, we must first label the roots, and because we plan to use `FPolynomialOrbit`, we take the ordering according to *Maple* or *Mathematica*: in *Maple*, $\alpha_1 \approx 4.14$, $\alpha_2 \approx 2.67i$, $\alpha_3 \approx -4.14$, $\alpha_4 \approx -2.67i$; in *Mathematica*, $\alpha_1 \approx -4.14$, $\alpha_2 \approx 4.14$, $\alpha_3 \approx -2.67i$, $\alpha_4 \approx 2.67i$. Then, by substituting approximations of these roots into the three elements of the orbit T of t, we find which image of t corresponds to the root in K, and then we note which subgroup of S_n among the three $D_{4,i}$ leaves that image invariant.

```
>    FPolynomialOrbit(X1*X2^2*X3^3 +
>       X2*X3^2*X4^3 + X3*X4^2*X1^3 +
>       X4*X1^2*X2^3 + X3*X2^2*X1^3 +
>       X2*X1^2*X4^3 + X1*X4^2*X3^3 +
>       X4*X3^2*X2^3, [X1, X2, X3, X4]);
```

$$[X1\ X2^2\ X3^3 + X2\ X3^2\ X4^3 +$$
$$X3\ X4^2\ X1^3 + X4\ X1^2\ X2^3 + X3\ X2^2\ X1^3 +$$
$$X2\ X1^2\ X4^3 + X1\ X4^2\ X3^3 + X2^3\ X3^2\ X4,$$
$$X2\ X1^2\ X3^3 + X1\ X3^2\ X4^3 + X2^3\ X3\ X4^2 +$$
$$X4\ X2^2\ X1^3 + X3\ X1^2\ X2^3 + X1\ X2^2\ X4^3 +$$
$$X2\ X3^3\ X4^2 + X1^3\ X3^2\ X4,$$
$$X1\ X3^2\ X2^3 + X2^2\ X3\ X4^3 + X2\ X4^2\ X1^3 +$$
$$X4\ X1^2\ X3^3 + X2\ X3^2\ X1^3 + X3\ X1^2\ X4^3 +$$
$$X1\ X4^2\ X2^3 + X2^2\ X3^3\ X4]$$

```
In[23]:= FPolynomialOrbit[X1 * X2^2 * X3^3+
         X2 * X3^2 * X4^3 + X3 * X4^2 * X1^3+
         X4 * X1^2 * X2^3 + X3 * X2^2 * X1^3+
         X2 * X1^2 * X4^3 + X1 * X4^2 * X3^3+
         X4 * X3^2 * X2^3, {X1, X2, X3, X4}]
```

$$Out[23]= \{X1^3\ X2^2\ X3 + X1\ X2^2\ X3^3 +$$
$$X1^2\ X2^3\ X4 + X2^3\ X3^2\ X4 + X1^3\ X3\ X4^2 +$$
$$X1\ X3^3\ X4^2 + X1^2\ X2\ X4^3 + X2\ X3^2\ X4^3,$$
$$X1^2\ X2^3\ X3 + X1^2\ X2\ X3^3 + X1^3\ X2^2\ X4 +$$
$$X1^3\ X3^2\ X4 + X2^3\ X3\ X4^2 + X2\ X3^3\ X4^2 +$$
$$X1\ X2^2\ X4^3 + X1\ X3^2\ X4^3,$$
$$X1^3\ X2\ X3^2 + X1\ X2^3\ X3^2 + X1^2\ X3^3\ X4 +$$
$$X2^2\ X3^3\ X4 + X1^3\ X2\ X4^2 + X1\ X2^3\ X4^2 +$$
$$X1^2\ X3\ X4^3 + X2^2\ X3\ X4^3\}$$

```
>    FPolynomialOrbit(
>       X1*X2^2*X3^3 + X2*X3^2*X4^3 +
>       X3*X4^2*X1^3 + X4*X1^2*X2^3 +
>       X3*X2^2*X1^3 + X2*X1^2*X4^3 +
>       X1*X4^2*X3^3 + X4*X3^2*X2^3,
>       X, X^4 - 10*X^2 - 122);
```

$$[4880., -2209.072208\ I, 2209.072208\ I]$$

```
In[24]:= FPolynomialOrbit[
         X1 * X2^2 * X3^3 + X2 * X3^2 * X4^3+
         X3 * X4^2 * X1^3 + X4 * X1^2 * X2^3+
         X3 * X2^2 * X1^3 + X2 * X1^2 * X4^3+
         X1 * X4^2 * X3^3 + X4 * X3^2 * X2^3,
         {X1, X2, X3, X4}, X^4 - 10X^2 - 122]
```

$$Out[24]= \{0. + 2209.07\ I, 0. - 2209.07\ I, 4880. + 0.\ I\}$$

Now two of these elements are clearly nonrational, leaving one the root in K: 4880. Therefore, with *Maple*'s labelling of the roots, the Galois group is $D_{4,2}$, while with *Mathematica*'s labelling of the roots, the Galois group is $D_{4,1}$. (If the root were in the first position, the group would be $D_{4,2}$; if in the second, $D_{4,3}$; if in the third, $D_{4,1}$.) To check, we execute `FGaloisGroup`.

```
>   FDeclareSplittingField(K,
>       X^4-10*X^2-122);
----Details of field K ----
    Algebraic Numbers: [r1, r2, r3, r4]
    Dimension over Q: 8
    Minimal Polynomials: [
        r1^4-10*r1^2-122,
        r1^2+r2^2-10, r3+r1, r4+r2]
    Root Approximations: [
        4.1381584857255214911,
        2.6691488630239679935*I,
        -4.1381584857255214911,
        -2.6691488630239679935*I]
----
```

```
In[25]:= FDeclareSplittingField[K,
            X^4 - 10X^2 - 122]
----Details of field K ----
    Algebraic Numbers: {r6, r7, r8, r9}
    Dimension over Q: 8
    Minimal Polynomials: {
        -122 - 10 * r6^2 + r6^4,
        r6 + r7, -10 + r6^2 + r8^2, r8 + r9}
    Root Approximations: {
        -4.1381584857255214911,
        4.1381584857255214911,
        -2.6691488630239679936 I,
        2.6691488630239679936 I}
----
```

```
>   FGaloisGroup(K);

    [[[r1, r2, r3, r4], [r1, r2, r3, r4]],
     [[r1, r2, r3, r4], [r1, r4, r3, r2]],
     [[r1, r2, r3, r4], [r2, r1, r4, r3]],
     [[r1, r2, r3, r4], [r2, r3, r4, r1]],
     [[r1, r2, r3, r4], [r3, r2, r1, r4]],
     [[r1, r2, r3, r4], [r3, r4, r1, r2]],
     [[r1, r2, r3, r4], [r4, r1, r2, r3]],
     [[r1, r2, r3, r4], [r4, r3, r2, r1]]]
```

```
In[26]:= FGaloisGroup[K]

Out[26]= {{{r6, r7, r8, r9}, {r6, r7, r8, r9}},
          {{r6, r7, r8, r9}, {r6, r7, r9, r8}},
          {{r6, r7, r8, r9}, {r7, r6, r8, r9}},
          {{r6, r7, r8, r9}, {r7, r6, r9, r8}},
          {{r6, r7, r8, r9}, {r8, r9, r6, r7}},
          {{r6, r7, r8, r9}, {r8, r9, r7, r6}},
          {{r6, r7, r8, r9}, {r9, r8, r6, r7}},
          {{r6, r7, r8, r9}, {r9, r8, r7, r6}}}
```

28.4. Factorization and Resolvents

The calculation of resolvents provides a good way to determine Galois groups of polynomials of degree 4, and of some polynomials of degree 5, for it is not too difficult to calculate a resolvent of an order 20 transitive subgroup of S_5. However, the calculation of a resolvent for an order 10 subgroup of S_5 is quite difficult. In this case, we may use the two methods – factorization over number fields of successively higher degree, and resolvents for subgroups of successively smaller order – together. If, using resolvents, it is determined that the Galois group lies inside a transitive subgroup of order 20 in S_5, then we could use `FDeclareSplittingField` to determine the size. The truth is that the calculation of Galois groups of polynomials is quite difficult. See [34] and [37] for further material.

29. Exercise Set 2

A.

29.1* Prove that conjugacy of subgroups of a group G is an equivalence relation.

29.2 The *dihedral group* D_4 of order 8 appears as a subgroup of S_4 in three ways: $D_{4,1} = \langle(1234), (24)\rangle$; $D_{4,2} = \langle(1342), (23)\rangle$; and $D_{4,3} = \langle(1423), (34)\rangle$. For each $D_{4,i} \subset S_4$, find a polynomial t_i with stabilizer $D_{4,i}$ and the corresponding orbit of t_i. Then express the coefficients of the resolvent in terms of the coefficients of a general polynomial p using symmetric polynomials. Can you choose the t_i in such a way that the resolvent is the same? (Don't use the same polynomials found in the text!)

29.3 Use a resolvent for A_3 to produce algebraic numbers α that generate $\mathrm{Fix}(A_3)$ over $\mathbb{Q}$ for the following polynomials p with Galois group S_3 and A_3.

(1) $X^3 + 3X + 1$;
(2) $X^3 + X^2 + 1$;
(3) $X^3 - 20$;
(4) $X^3 + 2X^2 - 5X + 1$.

Indicate α by its minimal polynomial, or, if requested, by an arithmetic combination of the roots of p.

29.4 Use a resolvent for $D_{4,2}$ to produce algebraic numbers α that generate $\mathrm{Fix}(D_{4,2})$ over $\mathbb{Q}$ for the following polynomials p with Galois group S_4 over $\mathbb{Q}$.

(1) $X^4 + 3X^2 + 5X + 5$;
(2) $X^4 + X^2 + X + 2$;
(3) $X^4 + X^2 + X + 3$.

Indicate α by its minimal polynomial, or, if requested, by an arithmetic combination of the roots of p.

29.5 Use a resolvent for A_4 to produce algebraic numbers α that generate $\text{Fix}(A_4 \cap \text{Gal}(L/\mathbb{Q}))$ over $\mathbb{Q}$ where L runs through the splitting fields of the following polynomials p.

(1) $X^4 - 4X^2 - 1$;

(2) $X^4 - 3$;

(3) $X^4 + 3X^2 + 1$.

29.6 Use a resolvent for A_3 to determine which of the following polynomials has a Galois group over $\mathbb{Q}$ isomorphic to A_3.

(1) $X^3 - 5X^2 + 2X + 1$;

(2) $X^3 - 2X^2 - X + 1$;

(3) $X^3 + 2X + 1$;

(4) $X^3 - 1$;

(5) $(X^2 - 2)(X - 1)$;

(6) $X^3 - 3X^2 + 3X - 1$.

When the Galois group is not isomorphic to A_3, describe the group. (Note: Not all of the polynomials are irreducible. What can we say in those cases?)

29.7 Use the resolvents given in the text for A_4, D_4, and C_4 to determine the isomorphism classes of the Galois groups over $\mathbb{Q}$ of the following polynomials.

(1) $X^4 + 3X^3 + 4X^2 + 5X + 6$;

(2) $X^4 - 8X^3 + 4X^2 + 48X + 16$;

(3) $(X^3 - 1)(X - 1)$;

(4) $(X^2 - 2)^2$;

(5) $X^4 + 2X^3 - 2X^2 - 6X - 3$.

(Note: Not all of the polynomials are irreducible. What can we say in those cases?)

29.8 If the factorization of the resolvent contains a linear factor with a multiplicity greater than 1, we cannot use the Resolvent Factorization Theorem to gain information about the Galois group. Using a $D_{4,2}$ resolvent in the text, show that this situation occurs with $X^4 - 13$.

B.

29.9* Let G be a permutation group on a finite number of symbols. Show that if two subgroups H_1 and H_2 are conjugate in G, then the permutations in H_2 may be obtained from those in H_1 by renaming the symbols in the permutations. For example, $\{(), (12)\}$

and $\{(), (23)\}$ are conjugate in S_3, and one subgroup can be obtained from the other by renaming $1 \mapsto 2, 2 \mapsto 3$, and $3 \mapsto 1$.

29.10* Prove the following variant of the Resolvent Factorization Theorem: Suppose K is a subfield of $\mathbb{C}$, L is a splitting field over K of a polynomial p of degree n with roots $\alpha_1, \ldots, \alpha_n$, and $G = \mathrm{Gal}(L/K)$ is the Galois group, viewed as a subgroup of S_n by virtue of its action on the roots. Let $t \in K[X_1, \ldots, X_n]$ be a polynomial with stabilizer $H \subset S_n$. Then

(1) if $s^{-1}Gs \subset H$ for all $s \in S_n$, then $f_{t,p}(X)$ factors into linear terms over K;

(2) if $f_{t,p}(X)$ factors into linear terms of multiplicity 1 over K, then $s^{-1}Gs \subset H$ for all $s \in S_n$.

29.11 Address the problem in Exercise 29.8 by finding a new Galois resolvent for D_4 that gives information about $X^4 - 13$. (Hint: Try a sum of images of $X_1 + X_1 X_2$.)

29.12 Address the problem in Exercise 29.8 by using a different polynomial with the same splitting field as $X^4 - 13$. (Hint: Let α be a root of $X^4 - 13$. Then if we choose $\beta \in \mathbb{Q}(\alpha)$ such that $\mathbb{Q}(\beta) = \mathbb{Q}(\alpha)$, then the splitting fields containing these must be the same (why?) and $m_\beta(X)$ is another polynomial that may be used in place of $X^4 - 13$ in considering the factorization of the resolvent (why?). Find a suitable $m_\beta(X)$ that factors appropriately.)

29.13 Suppose p is a polynomial over the field $\mathbb{Q}$ with splitting field L and Galois group S_4. Let $\mathbb{Q}(\alpha)$ be the subfield fixed by $D_{4,1}$ (or $D_{4,2}$ or $D_{4,3}$). Prove that the splitting field of $m_{\alpha,\mathbb{Q}}$ over $\mathbb{Q}$ has Galois group S_3.

CHAPTER SIX

Some Classical Topics

Knowing the Galois correspondence for subfields of $\mathbb{C}$ and some strategies for the determination of Galois groups of field extensions is, of course, just the beginning. It is of interest to determine the Galois groups of families of polynomials, to determine field extensions of a $\mathbb{Q}$ with a specified Galois group, to find generalizations of the Galois theory of subfields of $\mathbb{C}$ to arbitrary fields, and to apply the tools of Galois theory to solve problems further afield. In this chapter, we give introductions to several of these topics.

30. Roots of Unity and Cyclotomic Extensions

Perhaps the most basic polynomials are those of the form $X^n - 1$. In this section, we study their roots, called roots of unity, and the field extensions they generate. We first define roots of unity and cyclotomic extensions.

Definition 30.1 (Root of Unity, Primitive Root of Unity). A *root of unity* ω is a root, in the complex numbers, of a polynomial of the form $X^n - 1$ for some $n \in \mathbb{N}$. We say that ω is a *primitive root of unity of order n*, or a *primitive nth root*, if ω is a root of $X^n - 1$, but not of $X^m - 1$ for any $1 \leq m < n$.

Note that the definition of a root of unity ω is equivalent to saying that ω is an element of finite order in the multiplicative group $\mathbb{C}^*$ of $\mathbb{C}$, and it will be useful to view roots of unity in this way. You may wish to recall some results on finite cyclic groups, particularly concerning which elements of a cyclic group are generators of that group. Observe also that each root of unity ω is a primitive nth root of unity for some n; this n is the same as the order of ω in $\mathbb{C}^*$.

Definition 30.2 (Cyclotomic Extension). Let K be a subfield of $\mathbb{C}$. A *cyclotomic extension* of a field K is an extension L generated over K by a root of unity ω: $L = K(\omega)$.

We seek to understand the field extensions $K(\omega)/K$. First, observe that every cyclotomic extension is in fact a splitting field extension, as follows. If ω is a primitive nth root of unity over K, then the cyclic subgroup of $K(\omega)^*$ generated by ω, namely $\{1, \omega, \ldots, \omega^{n-1}\}$, contains n distinct elements; if not, $\omega^i = \omega^j$ with $0 \le i, j < n$ implies $\omega^{|i-j|} = 1$ when $|i - j| < n$, which is a contradiction unless $i - j = 0$. Furthermore, each element in $\{1, \omega, \ldots, \omega^{n-1}\}$ is a root of $X^n - 1$. We deduce that $\{1, \omega, \ldots, \omega^{n-1}\}$ is the complete set of roots of $X^n - 1$, and each is contained in $K(\omega)$. Hence $K(\omega)/K$ is the splitting field of $X^n - 1$, where n is the order of ω.

Since cyclotomic extensions are splitting field extensions, we naturally consider their Galois groups. Our first result shows that we can capture $\mathrm{Gal}(K(\omega)/K)$ as a subgroup of a certain abelian group, $U(n)$, by which we denote the units (elements with a multiplicative inverse) of the commutative ring $\mathbb{Z}/n\mathbb{Z}$.

Theorem 30.3. *Let K be a subfield of $\mathbb{C}$, ω a primitive nth root of unity over K, and $L = K(\omega)$. Then $\mathrm{Gal}(L/K)$ is isomorphic to a subgroup of $U(n)$.*

Proof. Let $g \in \mathrm{Gal}(L/K)$. Then $g(\omega)$ is another root of $m_{\omega,K}$, which must divide $X^n - 1$ (why?); hence $g(\omega)$ is another root of $X^n - 1$. By the discussion above, the roots of $X^n - 1$ are precisely $\{1, \omega, \ldots, \omega^{n-1}\}$, and hence $g(\omega) = \omega^i$ for some $i \in \mathbb{Z}/n\mathbb{Z}$. Now the same argument for $g^{-1} \in \mathrm{Gal}(L/K)$ shows that $g^{-1}(\omega) = \omega^j$ for some $j \in \mathbb{Z}/n\mathbb{Z}$. Since gg^{-1} is the identity element of $\mathrm{Gal}(L/K)$,

$$\omega = gg^{-1}(\omega) = g(\omega^j) = (g(\omega))^j = (\omega^i)^j = \omega^{ij},$$

or $ij \equiv 1 \pmod{n}$. Hence $i \in U(n)$.

One checks that this association $g \mapsto i$ is in fact a homomorphism of groups from $\mathrm{Gal}(L/K)$ to $U(n)$: define $\varphi: \mathrm{Gal}(L/K) \to U(n)$ by $\varphi(g) = i$ in $0 \le i < n$ such that $g(\omega) = \omega^i$. Moreover, it is one-to-one: if $g(\omega) = \omega$ then g elementwise fixes every element of L and hence must be the identity of $\mathrm{Gal}(L/K)$. $\square$

Now one may expect that when the base field K is the field $\mathbb{Q}$ of rational numbers, the Galois group ought to be the entire group $U(n)$, and such an intuition would be correct. In order to approach that result, we introduce cyclotomic polynomials, the minimal polynomials over $\mathbb{Q}$ of roots of unity. (Note that $X^n - 1$ cannot be the minimum polynomial of any nth root of unity for $n > 1$: $X - 1$ is a factor!)

Definition 30.4 (Cyclotomic Polynomial). Let ω be a primitive nth root of unity. The *nth cyclotomic polynomial* $Q_n(X)$ is

$$Q_n(X) = \prod_{i \in U(n)} (X - \omega^i).$$

The polynomial $Q_n(X)$ does not depend on the choice of a primitive nth root ω of unity, as follows. Given a fixed primitive nth root of unity ω, every other primitive nth root of unity must be ω^i for some $i \in U(n)$ (why?). Now we show that each ω^i, $i \in U(n)$, is a primitive nth root of unity. Let k be the order of ω^i in $\mathbb{C}^*$. Since $(\omega^i)^k = 1$, $ik \equiv 0 \pmod{n}$. Because $i \in U(n)$ has a multiplicative inverse j in $\mathbb{Z}/n\mathbb{Z}$, $ik \equiv 0 \pmod{n}$ implies that $jik \equiv 0 \pmod{n}$, or $k \equiv 0 \pmod{n}$. Because $0 < k \leq n$, we deduce that $k = n$. As a result, ω^i is a primitive root of unity of order n. The foregoing shows that $Q_n(X)$ is in fact the product of all $X - \omega$, where ω ranges over all primitive nth roots of unity. As a result, $Q_n(X)$ does not depend on the choice of ω.

Next, we explain why $Q_n(X)$ is a polynomial with coefficients in $\mathbb{Q}$, that is, $Q_n(X) \in \mathbb{Q}[X]$. By Exercise 26.7, each coefficient of $Q_n(X) \in \mathbb{Q}(\omega)[X]$ is left unchanged by every permutation of the roots $\{\omega^i : i \in U(n)\}$. Since each $g \in \mathrm{Gal}(\mathbb{Q}(\omega)/\mathbb{Q})$ sends ω^i, $i \in U(n)$, to ω^{ik}, $ik \in U(n)$, each coefficient lies in the fixed field of $\mathrm{Gal}(\mathbb{Q}(\omega)/\mathbb{Q})$. Then, by the Galois correspondence, each coefficient of $Q_n(X)$ lies in $\mathbb{Q}$.

It is immediate that $\deg Q_n(X) = |U(n)| = \phi(n)$ (where ϕ is the Euler phi function of n), and hence $Q_n(X) = m_{\omega,\mathbb{Q}}(X)$. Now in fact $Q_n(X) \in \mathbb{Z}[X]$, by Exercise 30.12. It is not immediate, however, that $Q_n(X)$ is irreducible.

Theorem 30.5. *$Q_n(X)$ is irreducible in $\mathbb{Q}[X]$.*

(For a proof, consult one of several standard texts: [1, Thm. 41], [3, Thm. 33.3], [4, Thm. 6.5.5], [16, Thm. 15.3], [21, Thm. II.7.7], [23, Thm. 10.2.8], or [25, Cor. 21.6].)

Putting together the theorems, however, we have

Corollary 30.6. *Let ω be a primitive nth root of unity and $L = \mathbb{Q}(\omega)$. Then $\mathrm{Gal}(L/\mathbb{Q}) \cong U(n)$.*

Exercises

In the exercises, ω_n denotes a primitive nth root of unity. As needed, use `AlgFields` to explore splitting fields and minimal polynomials. Note that since we already know the Galois groups, there is no need to use `FDeclareSplittingField` to set up the data structures for $\mathbb{Q}(\omega)$ as a splitting field (which is computationally expensive); to explore a cyclotomic field, simply use `FDeclareField` to declare a field obtained by adjoining a single primitive nth root and determine if your questions in that context can be answered by factoring polynomials over this extension field.

When asked to determine a Galois group, however, give as useful a description of it as possible, not simply a list of permutations; similarly, when asked to determine a subfield, provide a generating algebraic number and its minimal polynomial over $\mathbb{Q}$.

30.1 Determine Q_n for $n = 1, 2, \ldots, 10$. Note that $Q_n(X) \mid X^n - 1$.

30.2 Show that $X^n - 1 = \prod_{d \mid n} Q_d(X)$ by gathering roots of unity of the same order together.

30.3 Let p be a prime. Show that $Q_{p^n}(X) = (X^{p^n} - 1)/(X^{p^{n-1}} - 1)$.

30.4 Show that $Q_{np}(X) = Q_n(X^p)$ for $p \mid n$.

30.5 Determine the Galois groups of the following.

 (1) $\mathbb{Q}(\omega_3)/\mathbb{Q}$;
 (2) $\mathbb{Q}(\omega_9)/\mathbb{Q}$;
 (3) $\mathbb{Q}(\omega_8)/\mathbb{Q}$;
 (4) $\mathbb{Q}(\omega_8)/\mathbb{Q}(\omega_4)$;
 (5) $\mathbb{Q}(\omega_9)/\mathbb{Q}(\omega_3)$.

30.6 For $n > 2$, show that $[\mathbb{Q}(\omega_n + \omega_n^{-1}) : \mathbb{Q}] = \phi(n)/2$. (Hint: Find $m_{\omega_n, \mathbb{Q}(\omega_n + \omega_n^{-1})}$.)

30.7 Determine all intermediate fields of the following.

 (1) $\mathbb{Q}(\omega_8)/\mathbb{Q}$;
 (2) $\mathbb{Q}(\omega_{11})/\mathbb{Q}$;
 (3) $\mathbb{Q}(\omega_{12})/\mathbb{Q}$;
 (4) $\mathbb{Q}(\omega_{13})/\mathbb{Q}$.

(Hint: Consider expressions like $\omega + \omega^{-1}$. What elements of the Galois group fix such an expression?)

30.8 Use Galois groups to prove that $\mathbb{Q}(\sqrt[3]{5}) \not\subset \mathbb{Q}(\omega_n)$ for any n.

30.9 Prove that if m and n are relatively prime, then $U(mn) \cong U(m) \oplus U(n)$. (Hint: Consider the natural map from $\mathbb{Z}/mn\mathbb{Z}$ to $\mathbb{Z}/m\mathbb{Z} \oplus \mathbb{Z}/n\mathbb{Z}$.)

30.10 Use the previous exercise and the facts that $U(p^n) \cong \mathbb{Z}/(p^n - p^{n-1})\mathbb{Z}$ for $p > 2$ and $U(2^n) \cong \mathbb{Z}_2 \oplus \mathbb{Z}/2^{n-2}\mathbb{Z}$ for $n \geq 3$ to show that for any m, there exists an extension $L/\mathbb{Q}$ with Galois group isomorphic to $\mathbb{Z}/m\mathbb{Z}$. (Hint: Choose n carefully so that the decomposition of $U(n)$ has a factor group isomorphic to $\mathbb{Z}/m\mathbb{Z}$.)

30.11 Use the previous exercise and the idea of Exercise 30.6 to find an α such that $\mathbb{Q}(\alpha)/\mathbb{Q}$ has a cyclic Galois group of order 8. Verify computationally that the Galois group is $\mathbb{Z}/8\mathbb{Z}$, and find $m_{\alpha,\mathbb{Q}}(X)$.

30.12* Use the Division Algorithm for Integral Domains (Exercise 5.14) and Exercise 30.2 to prove that $\mathbb{Q}_n(X) \in \mathbb{Z}[X]$ for every n.

31. Cyclic Extensions over Fields with Roots of Unity

For each $n \in \mathbb{N}$, there exist field extensions L over $\mathbb{Q}$ with Galois group $\mathrm{Gal}(L/\mathbb{Q})$ isomorphic to $\mathbb{Z}/n\mathbb{Z}$ (see Exercise 30.10). However, for a fixed n, it is still difficult to understand the family of all such extensions.[1]

Over a subfield K of $\mathbb{C}$ containing a primitive nth root of unity, the situation is much different. Given $n \in \mathbb{N}$, the family of all extensions L of K with $\mathrm{Gal}(L/K) \cong \mathbb{Z}/n\mathbb{Z}$ is much more easily described: it is a subset of all extensions $K(\sqrt[n]{c})$ generated by an nth root over K. Reaching this result is the content of this section.

We begin with the simpler direction. For $c \in K^*$, denote by $\bar{c}$ the coset cK^{*n} in K^*/K^{*n}.

Theorem 31.1. *Let K be a subfield of $\mathbb{C}$ containing a primitive nth root of unity ω, and let L be the splitting field of $p(X) = X^n - c \in K[X]$ over K, with $c \neq 0$. Then $\mathrm{Gal}(L/K)$ is a cyclic group of order $d \mid n$, where d is the order of $\bar{c}$ in K^*/K^{*n}.*

Note that d might not equal n; while $\mathbb{Q}(\sqrt{c})$ may appear always to give a splitting field over $\mathbb{Q}$ with Galois group isomorphic to $\mathbb{Z}/2\mathbb{Z}$, setting $c = 4$ yields $\mathbb{Q}$ itself.

Proof. Let $\alpha_1, \ldots, \alpha_n$ be the roots of p and set $\alpha = \alpha_1$. It is an exercise (31.1) to show that $\{\alpha_i/\alpha\}_{i=1}^n$ is a set of n distinct roots of $X^n - 1$; hence the set of roots of p may be written $\{\alpha, \omega\alpha, \ldots, \omega^{n-1}\alpha\}$. Since $\omega \in K$, $L = K(\alpha)$.

---

[1] For a research monograph surveying recent progress, see [30].

Now for any element $g \in \text{Gal}(L/K)$, $g(\alpha) = \omega^i \alpha$ for some $i \in \mathbb{Z}/n\mathbb{Z}$. One checks that this association $g \mapsto i$ is in fact a homomorphism of groups from $\text{Gal}(L/K)$ to $\mathbb{Z}/n\mathbb{Z}$. Moreover, it is one-to-one: if $g(\alpha) = \alpha$, then g fixes every element of L and hence must be the identity of $\text{Gal}(L/K)$. Therefore $\text{Gal}(L/K)$ is isomorphic to the image of this homomorphism. Since the image is a subgroup of a cyclic group, it is cyclic as well. Denote by d the order of this image. By Lagrange's Theorem, $d \mid n$.

We now determine the size of this subgroup in terms of the order of $\bar{c}$. Let d denote the order of $\text{Gal}(L/K)$, let $g \in \text{Gal}(L/K)$ denote a generator of the Galois group, and define i as above by $g(\alpha) = \omega^i \alpha$. Observe that

$$\prod_{j=0}^{d-1} g^j(\alpha) = \prod_{j=0}^{d-1} \omega^{ij} \alpha = (\omega^i)^{\sum_{j=0}^{d-1} j} \alpha^d. \tag{31.1}$$

Now since the left-hand side of equation (31.1) is the product of α and all of its conjugates, the left-hand side is fixed by every element of the Galois group and is therefore in K. Hence the right-hand side is in K, and since $\omega \in K$, we deduce that $\alpha^d \in K$. Since $c \neq 0$, we have that $\alpha \neq 0$, and so $\alpha^d \in K^*$. Write $b = \alpha^d$.

Now $\alpha^n \in K^*$; $\alpha^n = c$. Moreover, we have that $c = b^{n/d}$. Raising both sides to the dth power gives $c^d = b^n \in K^{*n}$. Hence the order of $\bar{c}$ in the factor group K^*/K^{*n} divides d.

Let the order of $\bar{c}$ be e. We have shown that $e \mid d$ and hence $e \leq d$. Now we show that $d \leq e$, after which we will be done. Since $\bar{c}^n = \bar{1}$ in K^*/K^{*n}, e is a divisor of n. Since $(\bar{c})^e = \bar{1}$, $c^e = k^n$ for some $k \in K^*$, or $\alpha^{ne} = k^n$. Taking nth roots of both sides, $\omega^j \alpha^e = k$ for some j, or $\alpha^e = k/\omega^j \in K^*$. But then α is a root of $X^e - (k/\omega^j) \in K[X]$. Now $d = |\text{Gal}(L/K)| = [L : K] = [K(\alpha) : K] = \deg m_{\alpha,K}$, and $\deg m_{\alpha,K} \leq e$. Hence $d \leq e$. $\square$

For the other direction, we need the following proposition, a special case of Dedekind's Lemma (Exercise 31.4).

Proposition 31.2 (Independence of Automorphisms). *Let K be a subfield of $\mathbb{C}$, L a splitting field over K, and $G = \text{Gal}(L/K)$ the Galois group. Then there do not exist elements $l_g \in L$, $g \in G$, not all zero, such that*

$$\sum_{g \in G} l_g g(x) = 0, \qquad \forall x \in L.$$

Proof. Consider the more specific claim that for a nonempty subset (not subgroup!) S of G, there do not exist $l_s \in L$, $s \in S$, not all zero, such that

$$\sum_{s \in S} l_s s(x) = 0, \qquad \forall x \in L. \tag{31.2}$$

We prove this claim by induction on the size of the subsets S. This claim for $S = G$ is equivalent to the proposition.

To prove the claim in the case $|S| = 1$, suppose that $S = \{s\}$ and $l_s s(x) = 0$ for all $x \in L$. Then setting $x = 1$, we have that $l_s = 0$ and we are done.

For the inductive step, assume the claim holds for all subsets S with $|S| = n$ and suppose that we have a subset $S' \subset G$ of size $n+1$ and there exist $l_{s'} \in L$, $s' \in S'$, not all zero, such that

$$\sum_{s' \in S'} l_{s'} s'(x) = 0, \qquad \forall x \in L.$$

If any $l_{s'} = 0$, then this statement is the equivalent to the analogous statement for the complement $S = S' \setminus \{s'\}$ of $\{s'\}$ in S. This S is a subset of G of size n, so by induction we are done. Otherwise, suppose that all $l_{s'} \neq 0$ and choose a particular $s' \in S'$ with $l_{s'} \neq 0$. Dividing through by $l_{s'}$, we may assume that $l_{s'} = 1$. Then let $S = S' \setminus \{s'\}$. Our statement is then

$$s'(x) + \sum_{s \in S} l_s s(x) = 0, \qquad \forall x \in L. \tag{31.3}$$

Now since s' is an automorphism of L distinct from the $s \in S$, there exist $\bar{s} \in S$ and $l \in L$ such that $s'(l) \neq \bar{s}(l)$. Clearly $l \neq 0$ and so $s'(l) \neq 0$. Replacing x by lx in our equation and using the fact that for all $g \in G$, $g(lx) = g(l)g(x)$, we have

$$s'(l)s'(x) + \sum_{s \in S} l_s s(l)s(x) = 0, \qquad \forall x \in L. \tag{31.4}$$

Now multiplying equation (31.3) by $s'(l)$ and then subtracting equation (31.4), we have

$$\sum_{s \in S} \left(s'(l)l_s - l_s s(l) \right) s(x) = 0, \qquad \forall x \in L.$$

But the coefficients of $s(x)$ are independent of x and so we have an equation of the type of equation (31.2). By induction, $l_s(s'(l) - s(l)) = 0$ for all $s \in S$. Since $s'(l) \neq \bar{s}(l)$, we must have that $l_{\bar{s}} = 0$, a contradiction. $\qquad \square$

Theorem 31.3. *Let K be a subfield of $\mathbb{C}$ containing a primitive nth root of unity ω and L/K be a splitting field extension such that $\mathrm{Gal}(L/K) \cong \mathbb{Z}/n\mathbb{Z}$. Then there exists $c \in K^*$ such that L is the splitting field of $X^n - c \in K[X]$.*

Proof. Let g be a generator of $\mathrm{Gal}(L/K)$. Our strategy is to detect an nth root in L by finding a nonzero element α such that $g(\alpha) = \omega\alpha$. Suppose such an $\alpha \in L$ exists. Then

$$g^i(\alpha^n) = (g^i(\alpha))^n = (\omega^i\alpha)^n = \alpha^n$$

for each i, so α^n lies in $\mathrm{Fix}(\mathrm{Gal}(L/K)) = K$. If we set $c = \alpha^n$, then L will contain all of the roots of $X^n - c$, namely $\{\alpha, \omega\alpha, \ldots, \omega^{n-1}\alpha\}$. Then $K(\alpha)$ is a splitting field of $X^n - c$ inside L. Now $g^i(\alpha) = \omega^i\alpha$, so the only element of $\mathrm{Gal}(L/K)$ which leaves $K(\alpha)$ fixed is the identity. But then $\mathrm{Gal}(L/K(\alpha)) = \{id\}$ and, by the Galois correspondence, $L = K(\alpha)$.

To find such an α, we naively form an expression in L which satisfies $g(\alpha) = \omega\alpha$. We start with $x \in L$, add $\omega^{-1}g(x)$ and so on:

$$\alpha = x + \omega^{-1}g(x) + \omega^{-2}g^2(x) + \cdots + \omega^{-(n-1)}g^{n-1}(x).$$

We check that $g(\alpha) = \omega\alpha$, and we therefore need only that $\alpha \neq 0$. Note that the right-hand side is a linear combination

$$id(x) + \omega^{-1}g(x) + \cdots + \omega^{-(n-1)}g^{n-1}(x) = (id + \omega^{-1}g + \cdots + \omega^{-(n-1)}g^{n-1})(x)$$

of elements of the Galois group applied to x. Hence by Independence of Automorphisms (Proposition 31.2), there exists a nonzero x such that α is nonzero, and we are done. $\square$

Exercises

In the exercises, ω_n denotes a primitive nth root of unity. As needed, use `AlgFields` to explore splitting fields and minimal polynomials.

When asked to determine a Galois group, however, give as useful a description of it as possible, not simply a list of permutations.

31.1* Show that the n roots $\alpha_1, \ldots, \alpha_n$ of $X^n - c \in K[X]$ differ multiplicatively by roots of $X^n - 1$; show that $\{\alpha_i/\alpha_1\}_{i=1}^n$ is the complete set of roots of $X^n - 1$.

31.2 Determine the Galois groups of the splitting fields of the following polynomials.

(1) $X^2 - 2$ over $\mathbb{Q}$;

(2) $X^3 - 3$ over $\mathbb{Q}(\omega_3)$;

(3) $X^4 - 2$ over $\mathbb{Q}(i)$;

(4) $X^4 - 2$ over $\mathbb{Q}(\omega_8)$;

(5) $X^6 - 3$ over $\mathbb{Q}(\omega_3)$;

(6) $X^6 + 3$ over $\mathbb{Q}(\omega_3)$.

31.3 For the following field extensions $K(\beta)/K$ with cyclic Galois group, find $c \in K$ and $n \in \mathbb{N}$ such that $K(\beta)$ is the splitting field of $X^n - c$. (Hint: Pick an $x \in K(\beta)$ at random and work through the proof of Theorem 31.3, using `FDeclareSplittingExtensionField`, `FGaloisGroup`, and `FMinPoly`.)

(1) $K = \mathbb{Q}$, $m_{\beta,K}(X) = X^2 + 3X + 1$;

(2) $K = \mathbb{Q}(\omega_3)$, $m_{\beta,K}(X) = X^3 - 9X - 12$;

(3) $K = \mathbb{Q}(\omega_4)$, $m_{\beta,K}(X) = X^4 - 6X^2 - 12X + 6$.

31.4* [Dedekind's Lemma] Proposition 31.2 generalizes as follows. Let L be a field, H a group, and $f_1, \ldots, f_n \colon H \to L^*$ some distinct group homomorphisms. (Note that since L^* is abelian, each homomorphism has the commutator subgroup of H in the kernel.) Dedekind's Lemma states that there is no nontrivial linear combination of the $f_1, \ldots, f_n$ over L that is identically 0, that is, there do not exist $a_i \in L$, not all zero, such that $\sum a_i f_i(h) = 0$ for all $h \in H$. Follow the proof of the proposition to prove this result, noting that the proposition holds for the special case where H is L^* and the f_i are the elements of the Galois group $\mathrm{Gal}(L/K)$, restricted to the multiplicative subgroup of L.

31.5* Use Proposition 31.2 to prove that if G is a set – not necessarily a group – of automorphisms of a field L and L^G denotes the subfield consisting of elements of L fixed by every element of G, then $[L : L^G] \geq |G|$. (Sketch: Go by contradiction. Choose a basis $\{y_i\}$ of L over L^G, and consider the system of linear equations $\sum_{g \in G} g(y_i) X_g = 0$ with $|G|$ unknowns $X_g \in L$, for each i. Comparing the number of equations and unknowns, by linear algebra, the homogeneous system must have a nontrivial solution, that is, with not all $X_g = 0$. Now write an arbitrary $y \in L$ as a linear combination of the y_i over L^G and consider $\sum_{g \in G} g(y) X_g$. Show that this sum is zero and that this contradicts Proposition 31.2.)

31.6* Prove that if G is a group of automorphisms of a field L and L^G denotes the subfield consisting of elements of L fixed by every element of G, then $[L : L^G] \leq |G|$. (Sketch: Go by contradiction. Choose a basis $\{y_i\}$ of L over L^G, and consider the system S of linear equations $\sum_i g(y_i) X_i = 0$ with $[L : L^G]$ unknowns $X_i \in L$, for each $g \in G$. Comparing the number of equations and unknowns, by linear algebra, the homogeneous system must have a nontrivial solution, that is, with not all $X_i = 0$. Choose such a solution with the fewest nonzero X_i, and then, by renumbering the X_i if necessary, assume that the nonzero unknowns are $X_1, \ldots, X_k$. Now for $g' \in G$, note that the equations become $\sum_i g'g(y_i)g'(X_i) = 0$, or, equivalently, a system $T: \sum_i g(y_i)g'(X_i) = 0$, for each $g \in G$. Now multiply the equations in S by $g'(X_1)$ and the equations in T by X_1, and subtract corresponding equations for each $g \in G$. Now, as in the last exercise, derive a contradiction from Proposition 31.2 unless $X_i g'(X_1) - X_1 g'(X_i) = 0$ for each i and each $g' \in G$. Rewrite in terms of X_i/X_1. But then for each i, $X_i = l_i X_1$ for some $l_i \in L^G$. Use this fact in the original system S with $g = id$ to derive a contradiction.)

32. Binomial Equations

With an understanding, on one hand, of field extensions $K(\omega)/K$ for ω a primitive nth root of unity, and, on the other, of splitting fields of $X^n - c$ over fields containing a primitive root of unity of order n, we may now tackle the splitting fields of irreducible binomials $X^n - c$ over an arbitrary subfield K of $\mathbb{C}$. As we know, these are the extensions obtained by adjoining to K all the nth roots of the given c, and we will encounter both nth roots of unity and nth roots of c in studying these splitting fields.

Just as we showed that the Galois groups of cyclotomic extensions are isomorphic to subgroups of $U(n)$, we will show that the Galois groups of binomials are isomorphic to subgroups of a certain group, $(\mathbb{Z}/n\mathbb{Z}) \rtimes U(n)$.

Definition 32.1 (Semidirect Product). Given two groups A and B and a homomorphism $f: B \to \text{Aut}(A)$ from B to the group of automorphisms of A, the *semidirect product* $A \rtimes_f B$ is the set of pairs $\{(a, b) \mid a \in A, \ b \in B\}$ with the group operation

$$(a, b)(a', b') = \left(a \cdot (f(b))(a'), \ b \cdot b'\right),$$

where $\cdot$ represents the group operations in A and B, respectively.

Note that $f(b)$ is an automorphism of A, and hence $(f(b))(a')$ is another element of the group A. By abuse of language one often speaks of "the semidirect product $A \rtimes B$" when the homomorphism f is understood; without such a specified f, however, the notation $A \rtimes B$ is insufficient to define the group.

Now in our case, $U(n)$ acts naturally on $\mathbb{Z}/n\mathbb{Z}$ by multiplication, and we find that $U(n)$ is a group of automorphisms of $\mathbb{Z}/n\mathbb{Z}$, as follows. We can check that multiplication by $b \in U(n)$, given by the function

$$v_b \colon \mathbb{Z}/n\mathbb{Z} \to \mathbb{Z}/n\mathbb{Z}, \qquad v_b(a) = ab, \ \ a \in \mathbb{Z}/n\mathbb{Z}$$

is certainly a homomorphism of the additive group $\mathbb{Z}/n\mathbb{Z}$ to itself. Since multiplication by $b^{-1} \in U(n)$ is also such a homomorphism, we may compose the two homomorphisms to yield the identity. But then each homomorphism must in fact be an isomorphism. Hence $f \colon U(n) \to \mathrm{Aut}(\mathbb{Z}/n\mathbb{Z})$, given by $f(b) = v_b$, is a function, and we can check that f itself is in fact a homomorphism: $f(bb') = v_{bb'} = v_b \circ v_{b'} = f(b) \circ f(b')$.

We denote by $(\mathbb{Z}/n\mathbb{Z}) \rtimes U(n)$ the semidirect product $(\mathbb{Z}/n\mathbb{Z}) \rtimes_f U(n)$: the set of pairs (a, b), $a \in \mathbb{Z}/n\mathbb{Z}$, $b \in U(n)$, endowed with the group operation

$$(a, b)(a', b') = (a \cdot (f(b))(a'), \ b \cdot b') = (a + v_b(a'), \ bb') = (a + a'b, \ bb').$$

Clearly this group has order $n|U(n)| = n\phi(n)$ and its identity is $(0, 1)$.

We are now ready to consider the Galois groups of binomials:

Theorem 32.2. *Let K be a subfield of $\mathbb{C}$, $p(X) = X^n - c \in K[X]$ an irreducible polynomial over K, and L the splitting field of p over K.*

Then $\mathrm{Gal}(L/K)$ is isomorphic to a subgroup H of $(\mathbb{Z}/n\mathbb{Z}) \rtimes U(n)$.

Proof. Since the quotient of any two roots of p is an nth root of unity, and there are n distinct such quotients, we must have that ω, a primitive nth root of unity, lies in L.

Let $G = \mathrm{Gal}(L/K)$. For any $g \in G$, $g(\omega) = \omega^j$ for some $j \in \mathbb{Z}/n\mathbb{Z}$ since ω is a root of $X^n - 1$. By the same argument as in the beginning of the proof of Theorem 30.3, j must in fact be an element of $U(n)$. Let α be a root of $p(X)$. Then we may define $\varphi \colon G \to (\mathbb{Z}/n\mathbb{Z}) \rtimes U(n)$ by

$$\varphi(g) = (i, j)$$

where

$$g(\alpha) = \omega^i \alpha, \qquad g(\omega) = \omega^j.$$

It is an exercise (32.3) to check that φ is in fact a homomorphism. To show that it is one-to-one, suppose that $\varphi(g) = (0, 1)$. Since $i = 0$, $g(\alpha) = \omega^0 \alpha = \alpha$. Since $j = 1$, $g(\omega) = \omega$. Hence g fixes every element of L, so $g = id \in \mathrm{Gal}(L/K)$, and φ is one-to-one. If we let H be the image of φ, we are done. $\qquad\square$

Exercises

32.1 Prove that the semidirect product $A \times_f B$ satisfies the axioms of a group.

32.2 Prove that if $f : B \to \mathrm{Aut}(A)$ is given by $f(b) = id$, then $A \times_f B \cong A \times B$.

32.3* Prove that the function φ as defined in Theorem 32.2 is a homomorphism.

32.4 Find examples of polynomials $p \in \mathbb{Q}[X]$ with Galois groups isomorphic to each of the following:

    (1) $(\mathbb{Z}/2\mathbb{Z}) \rtimes U(2)$;

    (2) $(\mathbb{Z}/3\mathbb{Z}) \rtimes U(3)$;

    (3) $(\mathbb{Z}/4\mathbb{Z}) \rtimes U(4)$;

    (4) $(2\mathbb{Z}/4\mathbb{Z}) \rtimes U(4)$ (Hint: Try $X^4 - a^2$);

    (5) $(\mathbb{Z}/5\mathbb{Z}) \rtimes U(5)$;

    (6) $(\mathbb{Z}/6\mathbb{Z}) \rtimes U(6)$;

    (7) $(2\mathbb{Z}/6\mathbb{Z}) \rtimes U(6)$;

    (8) $(3\mathbb{Z}/6\mathbb{Z}) \rtimes U(6)$.

32.5 Suppose that K is a subfield of $\mathbb{C}$, n is a prime, a primitive nth root of unity does not lie in K, and $X^n - c$ is irreducible in $K[X]$. Prove that if L is the splitting field of $X^n - c$ over K then $\mathrm{Gal}(L/K) \cong \mathbb{Z}/n\mathbb{Z} \rtimes U(n)$. (Hint: First handle the case in which $X^n - c$ is irreducible over $K(\omega)$. Then show that if $X^n - c$ is irreducible over K, it is irreducible over $K(\omega)$, by showing that any factor of degree less than n of $X^n - c$ over $K(\omega)$ contains a coefficient of the form $s(\sqrt[n]{c})^m$ with $s \in K(\omega)^*$ and $1 \le m \le n-1$. Conclude that $\sqrt[n]{c} \in K(\omega) \setminus K$, and find a contradiction.)

33. Ruler-and-Compass Constructions

Ruler-and-compass construction problems go back to antiquity, and those that remained unsolved for centuries occupy a distinguished place in mathematical history. In this section, we consider the following four classical ruler-and-compass construction problems, showing how some elementary field theory – and for the last, a bit of Galois theory – resolves them:

(1) *Trisecting an Angle.* Whether, given an (arbitrary) angle, one may trisect that angle;
(2) *Doubling a Cube.* Whether, given an (arbitrary) segment, one may construct another segment such that the cube with sides congruent to the second segment has twice the volume of the cube with sides congruent to the first segment;
(3) *Squaring a Circle.* Whether, given (arbitrarily) a circle, one may construct a square with area equal to that of the circle; and
(4) *Constructing Regular Polygons.* Whether, given (arbitrarily) a positive integer $n \geq 3$, one may construct a regular n-gon, that is, a regular polygon with n sides.

First we settle on what geometric operations are allowed in a solution. Then we consider how these sorts of problems may be translated into field-theoretic questions about a particular field, the field of constructible numbers. Finally, we prove some results about constructible numbers that help us resolve the four problems.

33.1. Ground Rules and Basic Constructions

The Greeks assumed that the ruler and compass could be used to accomplish only two particular operations:

- The ruler may be used only as a straightedge, that is, to draw the line containing two points already given.
- The compass may be used only to draw the circle centered at a point already given and containing another point already given. (That is, after the drawing of one circle, the compass does not retain the length of the radial segment for use in drawing another circle.)

Now it turns out that these ground rules for construction problems are equivalent to ground rules that incorporate an additional operation: that of laying off, at a point on a line, a segment that is congruent to a segment given by two points. Permitting this additional operation is what is meant by using a "marked ruler" or a "noncollapsible compass." That the sets of ground rules are equivalent is an exercise (33.2), and we will assume for the remainder of the section that we are permitted precisely this expanded set of geometric operations.

In what follows, we will require three particular and basic Euclidean constructions. Take your compass out for a spin and try them out. We omit proofs that the constructions produce the desired geometric objects, but see [27] and [28] for much more in that vein.

Given a line l and a point P on l, construct a line through P perpendicular to l. Here we assume that we are given the line l by knowing at least two points on it; without loss of generality, then, Q is a point on l that is not P. Draw a circle with center P and radius PQ. The circle intersects l at another point R on the other side of P from Q. Now draw two circles, the first centered at Q and with radius QR, the second centered at R with radius RQ. These two circles intersect at two points, and the line through them is perpendicular to l and goes through P.

Given a line l and a point P not on l, construct a line through P perpendicular to l. Again, we assume that we are given l by knowing at least two points Q and R on it. Draw circles centered at P with radii PQ and PR. At least one of these circles must intersect l in two points. Call two such points S and T. Now draw two circles, the first centered at S and with radius ST, the second centered at T with radius TS. These two circles intersect at two points, and the line through them is perpendicular to l and goes through P.

Given a line l and a point P not on l, construct a line through P parallel to l. Use the second construction to find a line m perpendicular to l through P. Then use the first construction to find a line n perpendicular to m at P. The line n is parallel to l and goes through P.

33.2. From Geometry to Field Theory

We now consider how to translate construction problems into the language of field theory. Observe that construction problems generally ask whether, given some geometric objects at the outset like points, lines, or circles, one may use a ruler and a compass to create another geometric object with certain properties. (If no objects are given at the outset, we assume that we are given at least two points.) If, say, a point or a circle may be constructed using a ruler and compass according to the permissible operations, we will say that the point or circle is *constructible*.

We will translate the first three problems into field theory by first converting them into questions about lengths – more precisely, into questions about constructing a segment whose length is in a certain ratio to the length of another segment. Once we do so, we assign numbers to lengths of segments, and then the ratios become elements in a field.

To get started, we choose two particular given points P and Q (remember that we must have two to begin!) and declare as our unit of measurement the distance between P and Q. Then, we say that a real number d is *constructible* if, by a finite sequence of operations using a ruler and compass, we are able to construct two points A and B such that the distance between A and B is d units, or, in other words, that the ratio between the length of AB and the length of PQ is d.

Clearly 1 and 0 are constructible numbers. Moreover, some basic geometric constructions (Exercise 33.3) show that if two nonzero real numbers a and b are constructible, so are the numbers $a + b$, $a - b$, ab, and a/b. As a result, the set of constructible numbers is, quite fortunately, a field containing $\mathbb{Q}$. We denote this subfield of $\mathbb{R}$ by $\mathbb{K}$.

For the translation of the last problem, we will need an additional connection between angle measures θ of constructible angles and the constructibility, as real numbers, of $\sin \theta$ and $\cos \theta$, as follows:

If we are given three noncollinear points making an angle POQ with measure θ, the numbers $\sin \theta$ and $\cos \theta$ are constructible: lay off a unit segment on OQ beginning at O, ending at the constructed point R; construct the circle c centered at O containing R; label an intersection of c and the line through O and P as S; drop a perpendicular from S to the line through O and Q, thereby constructing a point of intersection T. Then the segment OT has length $\cos \theta$ and segment ST has length $\sin \theta$. It is an exercise (33.8) to show the converse, that if $\sin \theta$ and $\cos \theta$ are constructible numbers, then we may construct two lines such that an angle between them has angle measure θ. Hence we can convert questions about measures of angles that we construct to questions about ratios of lengths.

Now we have arrived at the translation of the four problems. In the language of constructible numbers, they are the following questions:

(1) Whether, given an arbitrary angle measure θ such that $\sin \theta$ and $\cos \theta$ are constructible numbers, the numbers $\sin(\theta/3)$ and $\cos(\theta/3)$ are constructible numbers;

(2) Whether $\sqrt[3]{2}$ is a constructible number;

(3) Whether $\sqrt{\pi}$ is a constructible number; and

(4) Whether $\sin(2\pi/n)$ and $\cos(2\pi/n)$ are constructible numbers. (Note that a regular n-gon has interior angles of measure $2\pi/n$.)

You should work out the details of why the four construction problems and the four questions are, respectively, equivalent.

33.3. From Field Theory to Impossibility Proofs

We develop some results on the field theory of constructible numbers in order to answer these questions.

Proposition 33.1. *If a real number c is constructible, then* $[\mathbb{Q}(c) : \mathbb{Q}]$ *is a power of 2.*

Proof. Recall that for construction problems, the initial unit of length is determined by two given points P and Q in the plane. Declare P to be the origin, the line l through P and Q to be the "x-axis," and the line m perpendicular to l at P to be the "y-axis." Doing so lays the groundwork for assigning coordinates to every constructible point in the plane: given such a constructible point R, drop perpendiculars to l, at X, and m, at Y, and consider the lengths of the segments PX and PY. These are the magnitudes of the x- and y-coordinates, and the signs may be determined by determining on which side of P, on their respective lines l and m, each of X and Y lies. Note that our construction yields coordinates (x, y) for a constructible point, and by the definition of constructible numbers, x and y are constructible. It is an exercise (33.4) to show that if x and y are constructible numbers, then the point R at coordinates (x, y) is constructible.

We have shown that determining which real numbers are the lengths of segments between constructible points is equivalent to determining which real numbers are coordinates of constructible points. Having done so allows us to consider the process of constructing points from previous ones in terms of equations.

A ruler-and-compass construction proceeds by constructing either a line from two previously constructed points, or a circle from a previously constructed point given as the center of the circle and another point given as lying on the circle. It is the intersections between pairs of these lines and circles that produce the constructible points. In terms of equations, what can we say about the constructible points?

Suppose we draw the line passing through two constructible points $P = (x, y)$ and $Q = (x', y')$. It is an exercise (33.5) to show that the equation of the line may be written in the form $aX + bY = c$ where $a, b, c \in \mathbb{K}$. Now suppose that we draw a circle with center a constructible point $P = (x, y)$ and a constructible point $Q = (x', y')$ on the circle. It is an exercise (33.6) to show that the equation of the circle may be written in the form $(X - x)^2 + (Y - y)^2 = c$, where $x, y, c \in \mathbb{K}$.

Now suppose the process of constructing a point requires finding, for instance, an intersection R of a line $aX + bY = c$ and a circle $(X - x')^2 + (Y - y')^2 = c'$, and we know that every scalar in these equations lies in $\mathbb{K}$. Assuming $b \neq 0$ and substituting

$Y = (c - aX)/b$ into the equation of the circle, we derive the equation

$$\left(1 + \frac{a^2}{b^2}\right)X^2 + \left(-\frac{2ac}{b^2} + \frac{2ay'}{b} - 2x'\right)X + \left(\frac{c^2}{b^2} - \frac{2cy'}{b} - c' + (x')^2 + (y')^2\right) = 0.$$

Observe that the coefficients lie in $\mathbb{K}$. Furthermore, the x-coordinate of the intersection point R is algebraic over $\mathbb{Q}(a, b, c, c', x', y')$ and is of degree at most 2. Since all of the coordinates are real numbers, we know from the quadratic formula that a solution X will be real so long as the expression under the square root sign in the quadratic formula is not negative.

It is an exercise (33.7) to carry on this analysis in the case of an intersection of a line and a circle when $b = 0$; the case of an intersection of a pair of lines; and in the case of an intersection of a pair of circles.

We conclude that each step in the process of constructing a point from previously constructed points results in coordinates that are constructible numbers of degree at most 2 over the field F generated over $\mathbb{Q}$ by the coordinates of the previously constructed points. By the Tower Law (Exercise 13.8), then, if c is a constructible number, then $[\mathbb{Q}(c) : \mathbb{Q}]$ is a power of 2.[2] $\square$

It is an exercise (33.1) to resolve the first three classical construction problems using Proposition 33.1. For the fourth, we need an additional result requiring Galois theory.

Proposition 33.2. *Let n be a positive integer. Then the following are equivalent:*

(1) $\phi(n) = 2^k$ for some k;
(2) $\cos(2\pi/n)$ is constructible;
(3) $\sin(2\pi/n)$ is constructible.

Proof. Assume first that (2) holds. We want to examine the degree $[\mathbb{Q}(\cos(2\pi/n)) : \mathbb{Q}]$, and we begin by observing that the field extension $\mathbb{Q}(\cos(2\pi/n))$ is a subextension of a cyclotomic extension $\mathbb{Q}(\omega)$, as follows. A primitive nth root of unity ω is given by $e^{2\pi i/n} = \cos(2\pi/n) + i\sin(2\pi/n)$. One checks that $\omega^{-1} = \cos(2\pi/n) - i\sin(2\pi/n)$ and that $(\omega + \omega^{-1})/2 = \cos(2\pi/n)$. Hence $\mathbb{Q}(\cos(2\pi/n)) \subset \mathbb{Q}(\omega)$. From Corollary 30.6, we know that $[\mathbb{Q}(\omega) : \mathbb{Q}] = |U(n)| = \phi(n)$.

Now (2) and Proposition 33.1 tell us that $[\mathbb{Q}(\cos(2\pi/n)) : \mathbb{Q}]$ is a power of 2. To use this statement to say something about $\phi(n)$, we will use the Tower Law (Exercise 13.8) once

---

[2] The converse to Proposition 33.1 is false. See [21, §III.16, Ex. 7] or [24, Cor. B.7, Remark].

we find $[\mathbb{Q}(\omega) : \mathbb{Q}(\cos(2\pi/n))]$. We find that this degree divides 2 as follows: ω is a root of

$$X^2 - 2\cos(2\pi/n)X + 1 \in \mathbb{Q}(\cos(2\pi/n))[X],$$

so that

$$[\mathbb{Q}(\omega) : \mathbb{Q}(\cos(2\pi/n))] \mid 2.$$

Hence (2) implies (1).

The identity $(\sin\theta)^2 + (\cos\theta)^2 = 1$ gives us that $\mathbb{Q}(\cos(2\pi/n))$ is contained in an algebraic extension of $\mathbb{Q}(\sin(2\pi/n))$ of degree at most 2. Hence, by the Tower Law (Exercise 13.8), (3) and Proposition 33.1 imply that $[\mathbb{Q}(\cos(2\pi/n)) : \mathbb{Q}]$ is a power of 2, and the preceding paragraph shows that (1) follows. Hence (3) implies (1).

Proceeding in the opposite direction, assume that (1) holds, that is, that $\phi(n)$ is a power of 2. As a field generated by a root of unity, $\mathbb{Q}(\omega)/\mathbb{Q}$ is then a field extension with abelian Galois group of order a power of 2 (Corollary 30.6). Since the subgroup $\mathrm{Gal}(\mathbb{Q}(\omega)/\mathbb{Q}(\cos(2\pi/n)))$ is normal, we may consider the Galois group G of the extension $\mathbb{Q}(\cos(2\pi/n))/\mathbb{Q}$, which must also be abelian of order a power of 2. It is an exercise (33.10) to show that every abelian group of even order contains a subgroup of index 2. Hence there exists a tower of subgroups

$$\{e\} = G_0 \subset G_1 \subset G_2 \subset \cdots \subset G_m = G$$

with successive subgroup indices $|G_{i+1} : G_i| = 2$. Let $F_i = \mathrm{Fix}(G_i)$. Then by the Galois correspondence we have a tower

$$F_0 = \mathbb{Q}(\cos(2\pi/n)) \supset F_1 \supset \cdots \supset F_m = \mathbb{Q}$$

of fields with successive degrees $[F_i : F_{i+1}] = 2$.

For subfields of $\mathbb{C}$, a field extension of degree 2 is generated by a square root: $F_i = F_{i+1}(\sqrt{a_{i+1}})$ for some $a_{i+1} \in F_{i+1}$. Since $\cos(2\pi/n) \in \mathbb{R}$ and each $F_i \subset \mathbb{Q}(\cos(2\pi/n)) \subset \mathbb{R}$, each $\sqrt{a_i} \in \mathbb{R}$. Now $a_m \in \mathbb{Q}$ and is therefore constructible. Moreover, given a constructible number a with $\sqrt{a} \in \mathbb{R}$, $\sqrt{a}$ is constructible (Exercise 33.9). Hence each $\sqrt{a_i}, i = m, \ldots, 1$, is constructible and we have that $\mathbb{Q}(\cos(2\pi/n))$ is a subfield of $\mathbb{K}$. Since $\mathbb{Q}(\sin(2\pi/n))$ is contained in an algebraic extension of $\mathbb{Q}(\cos(2\pi/n))$ of degree at most 2, the same argument shows that $\mathbb{Q}(\sin(2\pi/n))$ is a subfield of $\mathbb{K}$. Therefore $\sin(2\pi/n)$ and $\cos(2\pi/n)$ are constructible numbers, and we are done. $\qquad\square$

Hence we have resolved the fourth problem: a regular n-gon is constructible if and only if $\phi(n)$ is a power of 2.

Exercises

33.1* Use Proposition 33.1 to answer the first three classical construction problems negatively. (Hints: For the second, consider $m_{\sqrt[3]{2},\mathbb{Q}}(X)$. For the first, choose $\theta = \pi/3$ and use the triple-angle formula $\cos 3\theta = 4\cos^3 \theta - 3\cos \theta$ to arrive at $m_{\cos(\theta/3),\mathbb{Q}}(X) \mid 4X^3 - 3X - \cos\theta$. For the last, use the fact that π is transcendental.)

33.2* Prove that under the original requirements for ruler-and-compass constructions, it is possible to lay off a given segment on a given line at a given point. (Hint: Assume the line contains points P and Q and we seek to lay off on this line, beginning at P and in the same direction as Q, a segment AB. First construct a line l through A parallel to the line through P and Q. Then construct the circle c centered at A containing the point B. Label the intersection of l and c on the same side of the line AP as Q as C. Now construct the line m through C parallel to the line containing A and P. Label the intersection of m with the line through P and Q as D. Then the segment PD is congruent to AB.)

33.3* Prove that if a and b are nonzero constructible numbers, then $a + b$, $a - b$, ab, and a/b are also constructible. (Hint: For the first two, lay off segments in the appropriate directions; for the latter two, construct similar right triangles, one of which has legs of lengths a and 1, the other with a leg of length b.)

33.4* Prove that if x and y are constructible numbers, then a point R may be constructed such that the coordinates of R, as defined in the proof of Proposition 33.1, are (x, y).

33.5* Given two points $P = (x, y)$ and $Q = (x', y')$ with coordinates in $\mathbb{K}$, prove that the equation of the line passing through the points may be expressed in the form $aX + bY = c$, $a, b, c \in \mathbb{K}$. (Hint: First assume that the line is not vertical and write the equation first in the usual form $Y = mX + b$. Then generalize.)

33.6* Given two points $P = (x, y)$ and $Q = (x', y')$ with coordinates in $\mathbb{K}$, prove that the equation of the circle with center P and passing through Q may be expressed in the form $(X - a)^2 + (Y - b)^2 = c$ with $a, b, c \in \mathbb{K}$.

33.7* Given an intersection of two lines, a line and a circle, or two circles as in Exercises 33.5 and 33.6, prove that the coordinates of the intersection are algebraic of degree at most 2 over the field generated over $\mathbb{Q}$ by the coefficients of the two defining equations.

33.8* Prove that if $\sin\theta$ and $\cos\theta$ are constructible numbers, then lines l and m may be constructed such that an angle between them has an angle measure of θ.

33.9* Prove that if a is a constructible number with $\sqrt{a} \in \mathbb{R}$, then $\sqrt{a}$ is constructible. (Hint: Take a segment of length $a + 1$ and use bisection to construct a circle with this segment as a diameter. Now construct a point at length a from an endpoint of the original segment and erect a perpendicular to the segment at this point. Using the Pythagorean formula, find the length of the segment, on the perpendicular, from the line to the circle.)

33.10* The Fundamental Theorem of Finite Abelian Groups states that every finite abelian group is isomorphic to a direct sum of cyclic groups of prime-power order. Use this result to show that every abelian group of even order contains a subgroup of index 2.

34. Solvability by Radicals

One of the most exciting applications of basic Galois theory is the following statement:

> *There exist polynomials, of degree greater than 4 and with rational coefficients, whose roots cannot be expressed using any combination of the rational numbers, the four field operations of addition, subtraction, multiplication, and division, and the operation of taking integral roots.*

That is, there is no arithmetic combination involving rational numbers, square roots, cube roots, and so on, that will represent any root of such polynomials. This is a powerful claim, and once we determine that such polynomials exist with degree n, we may conclude that there is no analogue of the quadratic formula for polynomials of degree n. (However, there are analogues for cubic and quartic polynomials, though they are quite complicated!)

In order to formalize our claim, we introduce the notion of a radical extension and the notion of a polynomial's being solvable by radicals.

Definition 34.1. A *radical extension* K of a field F is a field that may be written $K = F(a_1, \ldots, a_m)$ with $a_i^{n_i} \in F(a_1, \ldots, a_{i-1})$ for $1 \le i \le m$ and $n_i \in \mathbb{N}$.

To say that an algebraic number a is in some radical extension of $\mathbb{Q}$, then, is equivalent to the statement that there is an arithmetic combination involving rational numbers, square roots, cube roots, and so on, that represents the algebraic number a. Note that there may be several ways of representing a field K as a radical extension of F, with different a_i and even different n_i, and so when we speak of a *particular representation* of a radical extension, we will mean the particular a_i and n_i used in satisfying the conditions of the definition above.

Definition 34.2. Let F be a subfield of $\mathbb{C}$ and $p \in F[X]$ an irreducible polynomial. Then p is *solvable by radicals* over F if each root a of p lies in a radical extension of F (which may depend on the root a).

Now the following lemma is crucial to our consideration of radical extensions, for it allows us to bring Galois theory to bear on the question of radical extensions:

Lemma 34.3. *Let F be a subfield of $\mathbb{C}$, and let K be a radical extension of F with a particular representation with $a_1, \ldots, a_m$ and n_i as in the definition. Let L be the splitting field over F of the product of minimal polynomials $m_{a_i, F}$. Then L is a radical extension of F with a particular representation*

$$L = F(b_1, \ldots, b_{m'}), \qquad b_i^{n_i'} \in F(b_1, \ldots, b_{i-1}), \ \ 1 \leq i \leq m'$$

such that the two sets $\{n_i\}$ and $\{n_i'\}$ are equal.

Proof. We prove the lemma by induction on m. If $m = 1$, then $a_1^{n_1} \in F$. Let $c = a_1^{n_1}$. Then $m_{a_1, F}(X)$ divides $X^{n_1} - c \in F[X]$, and if $\{b_1, \ldots, b_m\}$ is the set of roots of $m_{a_1, F}(X)$, then $b_i^{n_1} \in F$ for each i. Hence the splitting field L of $m_{a_1, F}(X)$ over F satisfies the conditions of the lemma.

Now we prepare for the inductive step. Suppose the lemma holds for $m - 1$, and let L be the splitting field of $\prod_{i=1}^{m} m_{a_i, F}$ over F. Let c be an arbitrary root of $m_{a_m, F}(X)$. There exists an isomorphism $f: F(a_m) \to F(c)$ (Theorem 18.6) over F, and we may extend this isomorphism to an automorphism f of L (Theorem 18.7).

Let $b_i = f(a_i)$ for each $1 \leq i \leq m$; then $c = b_m$. By properties of isomorphisms, $b_i^{n_i} \in F(b_1, \ldots, b_{i-1})$ for $1 \leq i \leq m$. Hence we have shown that each root c of $m_{a_m, F}(X)$ lies in a radical extension $F(b_1, \ldots, b_m)$, and the exponents n_i used in this particular representation of the radical extension $F(b_1, \ldots, b_m)$ are identical to the ones used in the particular representation of K over F. Having done so for each root c separately, now we must show that *all* roots lie in *one* radical extension with the desired exponents in a particular representation.

If we let d_{ij} be the b_i, and n_{ij} the n_i, used in the radical extension $F(b_1, \ldots, b_m)$ for the jth root of $m_{a_m, F}$, we see that adjoining all of the d_{ij} to F results in a radical extension of F containing all of the roots of $m_{a_m, F}$. (Note that for the set of exponents $\{n_{ij}\}$, i ranges from 1 to m and j ranges from 1 to $\deg m_{a_m, F}$.) For each j, these exponents $\{n_{ij}\}$ are identical to the set $\{n_i\}_{i=1}^m$ used in the particular representation of K over F. Hence the entire set $\{n_{ij}\}$ is equal to the set $\{n_i\}_{i=1}^m$.

Now let M be the splitting field over F of the product of polynomials $\prod_{i=1}^{m-1} m_{a_i, F}(X)$. Then L is the splitting field over M of the polynomial $m_{a_m, F}(X)$. By induction, we have that M is a radical extension of F, with a particular representation using some elements a_k, $k = 1, 2, \ldots, m'$, with exponents n'_k, such that $\{n_i\}_{i=1}^{m-1} = \{n'_k\}_{k=1}^{m'}$. Taking the union of this set of elements $\{a_k\}$ with our set $\{d_{ij}\}$, and observing that

$$\{n'_k\}_{k=1}^{m'} \cup \{n_{ij}\} = \{n_i\}_{i=1}^m,$$

we have a particular representation of L satisfying the conditions of the lemma. $\quad\square$

We derive an immediate consequence of this lemma. Suppose that a root a of an irreducible polynomial $p \in F[X]$ lies in a radical extension of K of F. Then the extension L of F given in the lemma is also a radical extension. Moreover, L contains all of the roots of p, by the second half of Theorem 19.1. Hence p is solvable by radicals over F. This shows that for an irreducible polynomial $p \in F[X]$, the following are equivalent:

(1) some root of p lies in a radical extension of F;
(2) p is solvable by radicals;
(3) there is a radical extension of F containing all of the roots of p.

Now we study the field extensions of the lemma, namely, splitting field extensions that are also radical extensions. We are particularly interested in what we can say about the Galois groups of such extensions.

Suppose L is a radical extension of F that is also a splitting field over F. Suppose that L has a particular representation as a radical extension of F using elements b_i, $i = 1, \ldots, m$, and corresponding exponents n_i. Set n to be the least common multiple of the n_i, and let ω be a primitive nth root of unity.

Since for each n_i there exists a d_i such that $n_i d_i = n$, we know that ω^{d_i} is a primitive n_ith root of unity. Now observe that $L(\omega)$ is a splitting field over F: if L is the splitting field over F of a polynomial $q \in F[X]$, $L(\omega)$ is the splitting field of $q(X) \cdot (X^n - 1)$.

To study our extension L/F, we study $L(\omega)/F$. Set $F_0 = F(\omega)$ and recursively define fields $F_i = F_{i-1}(b_i)$ for $i = 1, \ldots, m$. Then we have a tower of field extensions

$$F \subset F_0 = F(\omega) \subset F_1 = F_0(b_1) \subset \cdots \subset F_m = F_{m-1}(b_m) = L(\omega). \qquad (34.1)$$

Because $L(\omega)/F$ is a splitting field extension, we may let $G = \mathrm{Gal}(L(\omega)/F)$ and consider the various subextensions and their relation to the group G.

Generated by a root of unity, the extension F_0/F is a splitting field extension with an abelian Galois group (Theorem 30.3). Since each F_{i-1}, $1 \le i \le m$, contains a primitive n_ith root of unity, each F_i is a splitting field extension of F_{i-1} of a binomial $X^{n_i} - c_{i-1}$, for some $c_{i-1} \in F_{i-1}$, hence with an abelian Galois group (Theorem 31.1).

Using the Galois correspondence, we translate these conditions on subfields into conditions on subgroups. Let $G_i = \mathrm{Gal}(L(\omega)/F_i)$. We have a tower of subgroups

$$G \supset G_0 \supset G_1 \supset \cdots \supset G_{m-1} \supset G_m = \{e\},$$

with $G_i/G_{i+1} \cong \mathrm{Gal}(F_{i+1}/F_i)$ abelian for $i = 0, \ldots, m-1$, and G/G_0 abelian as well. (Take a moment to discover why G_{i+1} is normal in G_i.)

Renumbering our subgroups to include G, we abstract this property into a definition:

Definition 34.4. A finite group G is *solvable* if there exists a tower of subgroups

$$G = G_0 \supset G_1 \supset \cdots \supset G_{m-1} \supset G_m = \{e\}$$

such that G_{i+1} is normal in G_i and G_i/G_{i+1} is abelian, for all $0 \le i < m$.

Our analysis, then, has established the following:

Proposition 34.5. *Let F be a subfield of $\mathbb{C}$, and let L/F be an extension that is both a splitting field extension and a radical extension. Suppose L has a particular representation, as a radical extension, using exponents n_i, and let n be the least common multiple of the n_i and ω a primitive nth root of unity. Then $L(\omega)/F$ is a splitting field extension with a solvable Galois group.*

Now we proceed in the opposite direction, supposing first that a given splitting field extension has a solvable Galois group and concluding that the elements in the extension lie in some radical extension.

Recall that the *exponent* of a finite group G is the minimal positive integer n such that $g^n = e$ for all $g \in G$.

Proposition 34.6. *Let F be a subfield of* $\mathbb{C}$, *and let* L/F *be a splitting field extension with a solvable Galois group* $G = \mathrm{Gal}(L/F)$. *Let n be the exponent of G and* ω *a primitive nth root of unity. Then* $L(\omega)/F$ *is both a radical extension and a splitting field extension with solvable Galois group.*

Proof. Since G is solvable, there exists a tower of subgroups

$$G = G_0 \supset G_1 \supset \cdots \supset G_{m-1} \supset G_m = \{e\},$$

each subgroup normal in the preceding subgroup, such that G_i/G_{i+1} is abelian for all $i < m$. Let $K_i = \mathrm{Fix}(G_i)$. Then $\mathrm{Gal}(L/K_i) = G_i$, $\mathrm{Gal}(L/K_{i+1})$ is normal in $\mathrm{Gal}(L/K_i)$, and $\mathrm{Gal}(K_{i+1}/K_i) \cong G_i/G_{i+1}$ is abelian.

Now we consider another tower of fields, adjoining ω to each subfield: let $L_i = K_i(\omega)$ for all i. In particular, $L_m = L(\omega)$. Since L is the splitting field over F of some polynomial $p \in F[X]$ and L_m is the splitting field over F of $p(X) \cdot (X^n - 1)$, L_m/F is a splitting field extension. We denote by H its Galois group $\mathrm{Gal}(L_m/F)$.

Let $H_i = \mathrm{Gal}(L_m/L_i)$ for each i. By Exercise 26.14, L_{i+1} is a splitting field over L_i and $\mathrm{Gal}(L_{i+1}/L_i) \cong H_i/H_{i+1}$ is isomorphic to a subgroup of $\mathrm{Gal}(K_{i+1}/K_i) \cong G_i/G_{i+1}$, hence also abelian. Therefore we have a tower of subgroups

$$H_0 = \mathrm{Gal}(L_m/L_0) = \mathrm{Gal}(L_m/F(\omega))$$

$$\supset H_1 = \mathrm{Gal}(L_m/L_1)$$

$$\supset \cdots$$

$$\supset H_{m-1} = \mathrm{Gal}(L_m/L_{m-1})$$

$$\supset H_m = \{e\},$$

each subgroup normal in the preceding subgroup, such that H_i/H_{i+1} is abelian for all $i < m$. Now this tower does not begin with H; however, $H_0 = \mathrm{Gal}(L_m/L_0)$ is normal in $H = \mathrm{Gal}(L_m/F)$ since $L_0 = F(\omega)/F$ is a splitting field extension. Moreover, H/H_0 is abelian by Theorem 30.3. Hence L_m is a Galois extension of F with a solvable Galois group.

Now this tower is of a form identical to that of the tower (34.1), and we show that this similar form implies that L_m is a radical extension of F. First, $L_0 = F(\omega)$ is clearly a radical extension. It is an exercise (34.1) to show that, for each $i < m$, L_{i+1} is a radical extension of L_i. Hence L_m is a radical extension of F, and we are done. $\qquad\square$

With one more result, that quotients of solvable groups are solvable (Exercise 34.2), we may conclude with

Theorem 34.7. *Let F be a subfield of* $\mathbb{C}$. *Let* $p \in F[X]$ *be an irreducible polynomial and L the splitting field of p over F. Then the following are equivalent:*

(1) some root of p lies in a radical extension of F;
(2) there is a radical extension of F containing all the roots of p;
(3) p is solvable by radicals;
(4) Gal(L/F) *is a solvable group.*

Proof. We have already shown the equivalence of the first three items, and Proposition 34.6 gives us that the fourth item implies the first three.

Assume that L lies in a radical extension L' of F; this is equivalent to item (2). Then L' lies in a radical extension L'' of F that is also a splitting field extension (Lemma 34.3). Furthermore, there exists an ω such that $L''(\omega)/F$ is a radical extension that is a splitting field with solvable Galois group (Proposition 34.5). Since by Exercise 34.2, quotients of solvable groups are solvable, Gal(L/F) is solvable. Hence item (2) implies item (4) and all of the items are equivalent.   □

Now one may check that the polynomial $p(X) = X^5 + 20X + 16$ has Galois group A_5. In Exercise 34.4, we outline a proof that A_5 is not solvable. (In fact, A_5 is the smallest nonsolvable group.) We conclude, then, that p is not solvable by radicals. More generally, one can show that for every $n > 4$, there is a polynomial in $\mathbb{Q}[X]$ of degree n that is not solvable by radicals.

Exercises

34.1* Let K be a splitting field over F, and suppose that F contains ω, a primitive nth root of unity of order equal to the exponent of Gal(K/F). Suppose further that Gal(K/F) is abelian. Show that K/F is a radical extension. (Hint: Decompose $G = $ Gal(K/F) into a tower $G = G_0 \supset G_1 \supset \cdots \supset G_m = \{e\}$ of subgroups such that G_i/G_{i+1} is cyclic, using the Fundamental Theorem of Finite Cyclic Groups. Then use Theorem 31.3 repeatedly.)

34.2* Prove that a quotient of a solvable group is solvable. (Hint: Let $f: G \to G/M$ be the natural homomorphism onto a quotient group G/M, and consider a step $G_i \supset G_{i+1}$ in a tower. Show from the definition of normality that if G_{i+1} is normal in G_i, $f(G_{i+1})$ is normal in $f(G_i)$. Next show that if, as cosets, $(aG_{i+1})(bG_{i+1}) = abG_{i+1} = baG_{i+1}$, which expresses the abelian property for G_i/G_{i+1}, then using f, we can verify a similar relation for $f(G_i)/f(G_{i+1})$. Conclude that there exists a tower $\{f(G_i)\}$ satisfying the conditions of solvability.)

34.3 Prove that a subgroup of a solvable group is solvable.

34.4* Prove that A_5 is not solvable by consulting Appendix 2. (Hint: The subgroups of A_5 are those subgroups of S_5 in classes (1), (3), (4), (7), (8), (11), (13), (14), and (18). Note that conjugation pqp^{-1} of an n-cycle $q = (a_1 a_2 \cdots a_n)$ by a permutation p takes and "renames" the elements; the new cycle is $(p(a_1) p(a_2) \cdots p(a_n))$; see Exercise 29.9. Prove that for a subgroup in class (3), (7), (8), or (13), conjugation by an appropriate *even* permutation sends the subgroup to another subgroup in the same class. Hence no such nontrivial subgroups are normal, and each nontrivial normal subgroup of A_5 contains a 3-cycle. Now prove that conjugation by an appropriate *even* permutation sends a 3-cycle to any other 3-cycle. Hence each nontrivial normal subgroup of A_5 contains every 3-cycle. Finally show that every element of A_5 is a product of 3-cycles, by showing that a nondisjoint product of two 2-cycles is a 3-cycle, while a disjoint product of two 2-cycles may be expressed as a product of two 3-cycles. Hence the only nontrivial normal subgroup of A_5 is A_5 itself.)

34.5 Combine Exercises 34.3 and 34.4 to conclude that no alternating or symmetric group on at least five letters is solvable. Using the fact that for every n polynomials in $\mathbb{Q}[X]$ exist with Galois group S_n, conclude that there exist polynomials in $\mathbb{Q}[X]$ of every degree $n > 4$ that are not solvable by radicals.

35. Characteristic p and Arbitrary Fields

In this section, we consider Galois theory over arbitrary fields, not simply subfields of $\mathbb{C}$. In particular, we ask which of the logical connections we have built in this text may be generalized to this wider setting, as well as what concepts may be changed in order to generalize the rest.

Now arbitrary fields come in many flavors. Indeed, they may still be of characteristic zero yet not contained in $\mathbb{C}$. One example of such a field is $\mathbb{C}(t)$, the field of quotients of polynomials in one variable t with complex coefficients. On the other hand, arbitrary fields may be of prime characteristic p, and then either finite (as in the case of $\mathbb{Z}/p\mathbb{Z}$, often denoted $\mathbb{F}_p$) or infinite (as in the case of $\mathbb{F}_p(t)$). In this section, we will revisit, chapter by chapter, the theory leading to the Galois correspondence and examine what changes of perspective and of concepts are necessary to develop an attractive generalization.[3] We will not, however, consider how a symbolic computation system might be used to study examples of extensions of arbitrary fields; leaving number fields, we leave `AlgFields` as well.

35.1. Basic Algebra and $\mathbb{C}$

The first chapter of this text begins with material valid over arbitrary fields: notions of polynomials and polynomial rings, the Division Algorithm and the Euclidean Algorithm, and the notion of factorization. The Fundamental Theorem of Algebra (Theorem 1.13), however, marks the first place where our path must diverge, for we no longer assume that we have polynomials with coefficients that are elements of $\mathbb{C}$.

The Fundamental Theorem of Algebra asserts that given any polynomial p with complex coefficients, the field $\mathbb{C}$ of complex numbers contains a root of p. When many of us first encountered the Fundamental Theorem, we took its statement to be one of reassurance: once the real numbers are expanded to the complex numbers, each polynomial equation in one variable may be solved. Under this point of view, the complex numbers are the familiar object, while polynomials and their roots are the unfamiliar objects, and we were relieved to learn that polynomials have complete sets of roots inside the familiar system of the complex numbers.

It is precisely the opposite point of view that we now adopt. Given an arbitrary field K and $p \in K[X]$ an irreducible polynomial, we consider these objects to be *more* familiar than any extension field in which p might have a root. Indeed, given an arbitrary K, we may have no conception of what the extension fields of K are. Instead, we build up extension fields from the polynomials themselves: we assert that we may always find an extension field of K that is generated over K by an *element* that is a root of p. This is the content of Kronecker's Theorem:

Theorem 35.1. *(Kronecker, 1887 [53]) Let K be a field and p an irreducible polynomial. Then $L = K[X]/(p(X))$ is an extension field of K generated by an element that is a root of p.*

The proof is left as an exercise (35.1).

---

[3] A further direction one might take would be to generalize Galois theory to infinite-dimensional field extensions. We will not venture down this path, but treatments of the subject may be found in [19, §8.6] and [21, Chap. IV].

With Kronecker's Theorem, then, it is no surprise to us that a polynomial over a field has a root, in some extension field. In part because the theorem stands in for the Fundamental Theorem of Algebra, some call Kronecker's Theorem the Fundamental Theorem of Field Theory. What remains unclear, however, is how to understand these extensions of K containing roots of p, for we do not have a predetermined field such as the complex numbers in which the roots are assumed to lie.

Remark 35.2. We mention that there is a notion of an *algebraic closure* of a field K. Call a field L *algebraically closed* if it has the property that every polynomial in $L[X]$ has a root in L. An *algebraic closure* of a field K is an algebraic extension field L of K that is algebraically closed. This notion generalizes to L the statement about $\mathbb{C}$ in Theorem 1.13 (and even permits the same proof for Corollary 1.14). Said another way, the Fundamental Theorem of Algebra tells us that $\mathbb{C}$ is algebraically closed and hence contains an algebraic closure of any subfield K of $\mathbb{C}$. Steinitz proved in 1910 ([57]; see also [1, §13.4, Props. 29–31], [21, p. 30ff], and [23, §2.7]) that every field K has an algebraic closure and moreover that algebraic closures of K are unique up to isomorphism.

Note, however, that these notions do not serve to give us a well-understood substitute for $\mathbb{C}$. While $\mathbb{C}$ contains an algebraic closure of $\mathbb{Q}$, it is by no means the only algebraically closed field containing an algebraic closure of $\mathbb{Q}$. We denote by $\mathbb{Q}^{\mathrm{alg}}$ the algebraic closure of $\mathbb{Q}$ in $\mathbb{C}$; this field is simply the subfield of $\mathbb{C}$ consisting of algebraic numbers. The field $\mathbb{Q}^{\mathrm{alg}}$ is isomorphic, then, to any algebraic closure of $\mathbb{Q}$, but even knowing that it is unique up to isomorphism very likely leaves us no more familiar with $\mathbb{Q}^{\mathrm{alg}}$ than we were.

35.2. Algebraic Elements, Field Extensions, and Minimal Polynomials

Taking our new point of view into the second chapter, we find we must replace the notion of an algebraic number in $\mathbb{C}$ with that of an algebraic element in some extension field L of K. We have already encountered these ideas in Definition 19.6 (and in Exercise 10.22 as well). We therefore define an *algebraic element* α to be an element of an extension field L of K such that α is the root of some polynomial in $K[X]$. We say that α is then *algebraic over* K. Then an *algebraic extension* L of K is a field L containing K with the property that each element of L is algebraic over K. From Chapter 2 onwards, then, whenever we encounter an algebraic number α or a field $K(\alpha)$ generated over K by an algebraic number, we think instead of an algebraic element.

This new approach serves us well quite far into the second chapter. (See Exercise 35.2 for additional details of modifications required of the results there.) Note in particular that the natural generalization of the Structure Theorem for Fields Generated by an Algebraic Number (8.12) – that $K(\alpha)$ for α algebraic is isomorphic to $K[X]/(m_{\alpha,K}(X))$ – takes on an added significance, for while we may not have a grasp of all of the extension fields L of an arbitrary field K, we can say that all generated fields $K(\alpha) \subset L$ for an algebraic element α with minimal polynomial $m_{\alpha,K}$ are isomorphic to a quotient of a polynomial ring, namely $K[X]/(m_{\alpha,K}(X))$.

35.3. Working with Algebraic Elements: Separability

Moving on to the third chapter, we encounter a new difficulty: Theorems 11.2 and 11.3, which together tell us that every irreducible polynomial over a subfield K of $\mathbb{C}$ factors into distinct linear factors. In other words, an irreducible polynomial of degree n over such a field K has n roots of multiplicity 1. Without a familiar field such as $\mathbb{C}$ over which to factor polynomials into linear factors, we must decide in the arbitrary field case whether irreducible polynomials factor into distinct linear factors – that is, so that the roots have multiplicity 1 – in some field extension. The answer is contained in the following result.

Theorem 35.3. *Let K be a field, $f \in K[X]$ an irreducible polynomial, and L an extension field of K over which f factors into linear terms:*

$$f(X) = c \prod (X - \alpha_i), \ \ c \in L^*, \ \alpha_i \in L, \ i = 1, \ldots, \deg p.$$

Then the α_i are distinct, except in the case where K is a field of characteristic $p > 0$ and $f(X) = g(X^p)$ for some polynomial $g \in K[X]$. In this case, the multiplicity of each root α is divisible by p.

It is interesting to note that although the theorem does not show the existence of an extension field in which the polynomial factors into linear terms, it does tell us that, by examining the polynomial alone, we may decide whether the roots will be distinct. Ideas for a proof are contained in an exercise (35.3).

Faced with this theorem, we now make the standard definition of that property of polynomials that insures that our polynomials have only distinct roots:

Definition 35.4 (Separability). Let K be a field. If p is an irreducible polynomial of degree n, we say that p is *separable over K* if, in each extension field L of K over which p factors into linear terms, p has n distinct roots in L. If p is a nonconstant polynomial, we say that p is separable over K if all of its irreducible factors are separable over K. If α is an algebraic element in an extension field of K, we say that α is separable over K if its minimum polynomial over K is separable. We say that a field extension L/K is *separable* if every element of L is separable over K.

Of course, not every element of an extension field will be separable (Exercise 35.5).

To generalize Galois theory to arbitrary fields, we constrain ourselves to consider only those minimal polynomials that are separable. As a result, every algebraic element and every algebraic extension we consider are separable. Exercises 35.4 and 35.6 show that over fields of characteristic 0 and over finite fields, this condition is always satisfied. This requirement serves us well through the third chapter, though we do not yet know that a field extension generated over K by a separable algebraic element α is a separable extension; this result will come later. (See Exercise 35.7 for additional details of modifications required of results in the third chapter.)

35.4. Multiply Generated Fields and the Galois Correspondence

The fourth chapter considers fields generated by more than one algebraic number over a subfield of $\mathbb{C}$, and to generalize these results, we must grapple again with the fact that, having left the familiar realm of algebraic numbers, we have lost the notion of specifying a particular element of a field by giving a minimal polynomial and a complex approximation (or, instead, a "root number," as in section 14). In particular, we must recognize that to write $K(\alpha)(\beta)$ for algebraic elements α and β, we must be able to specify not only the minimal polynomial of α over K, which determines $K(\alpha)$ up to isomorphism, but also the minimal polynomial of β over $K(\alpha)$. This fact complicates matters when we consider how to define a splitting field.

Given a field K and a nonconstant polynomial $p \in K[X]$, we define a splitting field in a way that is natural for the arbitrary field case:

Definition 35.5 (Splitting Field). If K is a field and $p \in K[X]$ a nonconstant polynomial, then an extension field L of K is a *splitting field*, or L/K is a *splitting field extension*, if L is a minimal field extension with respect to the property that p factors into linear factors over L.

Minimal here means that, with respect to inclusion, there is no smaller extension with the same property. To validate such a definition, one must first verify that there exists at least one extension with the desired property. Intuitively, the idea is to continue to factor

p over successively larger fields, adjoining a root of an irreducible factor each time. The process must stop because a field may have only deg p roots of a polynomial p. Providing a rigorous proof is Exercise 35.8. Then, to preserve our intuition about the concept, one shows that splitting fields are, in fact, unique up to isomorphism (Exercise 35.9).

Now we seek a generalization of Theorem 19.1 – that adjoining an arbitrary number of algebraic elements results in an algebraic extension, and if further the complete set of roots of the minimal polynomial of each adjoined element is adjoined, the result is a normal extension – to incorporate separability. To do so, we will generalize some other results in an order different from the order in which they first appeared in the text.

First we adopt the same definition of a Galois group (24.1) as before, so long as L/K is the splitting field extension of a separable polynomial, and we adopt the second definition of normal (23.2) presented in Chapter 5.

Then we prove a generalization of the theorem that the order of a Galois group is equal to the dimension of the extension (Theorem 24.4):

Theorem 35.6. *Let K be a field, $p \in K[X]$ a separable polynomial, and L a splitting field for p over K. Then the number of automorphisms of L over K is equal to $[L : K]$. Hence $|\mathrm{Gal}(L/K)| = [L : K]$.*

Proof. Our strategy is to prove a more general statement: given fields K and K' isomorphic under an isomorphism $\tau : K \to K'$, a separable polynomial $p \in K[X]$, and splitting fields L of p over K and L' of $\tau(p)$ over K', there are precisely $[L : K]$ isomorphisms from L to L' extending τ. The theorem is then the special case in which $K' = K$, $L' = L$, and $\tau = id$.

We prove this more general statement by induction on $n = [L : K]$. The base case $n = 1$ is trivial. Now assume that $n > 1$ and the statement holds for all L/K with $[L : K] < n$. Factor p over K as $p = f_1 f_2$, where f_1 is irreducible of degree d, for some $1 < d \leq n$ (and f_2 may be constant). Let α be a root in L of $f_1(X)$.

Now any monomorphism from $K(\alpha)$ to L' must send α to a root of $\tau(f_1)$. Moreover, by a straightforward generalization of Theorem 18.6, given any $\alpha' \in L'$ that is a root of $\tau(f_1)$, there is an isomorphism from $K(\alpha)$ to $K'(\alpha') \subset L'$ extending τ. Hence the number of extensions of τ to monomorphisms $\bar{\tau}$ from $K(\alpha)$ to L' is equal to the number of roots of $\tau(p)$ in L'.

Separability is preserved under isomorphisms (Exercise 35.13), so $\tau(p)$ is separable over K'. Hence there exist $d = \deg \tau(p)$ roots of $\tau(p)$ in L', and we have d extensions $\bar{\tau}$ of τ to monomorphisms from $K(\alpha)$ to L'.

Now note that $L/K(\alpha)$ is a splitting field extension of a separable polynomial, namely p. Moreover, for any α' a root of $\tau(f_1)$, $L'/K'(\alpha')$ is similarly a splitting field of a separable polynomial, namely $\tau(p)$. Any extension $\bar{\tau}$ of τ to a monomorphism from $K(\alpha)$ to L' is an isomorphism from $K(\alpha)$ to some $K'(\alpha')$. Hence we may invoke the inductive hypothesis for $\tau \colon K(\alpha) \to K'(\alpha')$ and p to say that there are $[L : K(\alpha)] = [L : K]/d$ extensions of each $\bar{\tau}$ to isomorphisms from L to L'. Therefore there are $([L : K]/d) \cdot d = [L : K]$ extensions of τ to isomorphisms from L to L'. □

Then we may prove the following:

Theorem 35.7. *Let K be a field, $p \in K[X]$ a separable polynomial, and L a splitting field for p over K. Then every element $\alpha \in L$, and hence every intermediate field M, $K \subset M \subset L$, is separable over K.*

Proof. Let $G = \mathrm{Gal}(L/K)$, but note that while we have a definition of a Galois group, we do not yet have results about the Galois correspondence. By the previous theorem (35.6), $|G| = [L : K]$. Moreover, by Exercises 31.5 and 31.6, $|G| = [L : L^G]$, where $L^G = \{l \in L \mid g(l) = l,\ \forall g \in G\}$. Since $K \subset L^G$ and L has the same degree over each, $K = L^G$.

Now suppose $\alpha \in L$. Consider $\{g(\alpha)\}_{g \in G}$, the set of images of α under $g \in G$, and enumerate the distinct elements of this set $\alpha_1, \ldots, \alpha_m$. Form the polynomial $f(X) = \prod_{i=1}^{m}(X - \alpha_i)$. Now each $g \in G$ permutes the elements $\alpha_1, \ldots, \alpha_m$, and by Exercise 26.7, the coefficients of f are therefore left unchanged by any $g \in G$. Hence each coefficient of f lies in $L^G = K$, so $f(X) \in K[X]$. Moreover, f has distinct roots in L by construction.

Since L is a splitting field over K of a polynomial, $[L : K]$ is finite and hence α is algebraic over K. Now $m_{\alpha,K}(X)$, the minimal polynomial of α over K, must divide f. Hence, over L, $m_{\alpha,K}(X)$ must factor into distinct linear factors – all of the form $X - \alpha_i$ – and so α is separable. □

Using this result, we may go back and show that adjoining complete sets of roots of separable minimal polynomials of algebraic elements results in a normal extension. Doing so – generalizing Theorem 19.1 and the exercise generalizing it (20.18) – is an exercise (35.10).

Now one may generalize the Finite = Simple and Algebraic Theorem (19.7) and the exercise generalizing it (20.20) in the following way: Let L/K be a separable extension.

Then L/K is finite if and only if it is simple and algebraic. (See Exercise 35.12 for additional details of modifications required of results in the fourth chapter.)

With these definitions and theorems, the results of the fifth chapter will hold, so long as we restrict ourselves to splitting fields of separable polynomials. These particular fields earn a special definition:

Definition 35.8 (Galois extension). A finite field extension K/F is *Galois* if it is normal and separable.

Beginning with Theorem 23.4, we assume that all extensions are separable and all splitting fields are splitting fields of separable polynomials. In particular, the Galois correspondence holds for splitting fields of separable polynomials. Working through these proofs is Exercise 35.14.

Exercises

35.1* Prove Kronecker's Theorem: Show how K embeds in L so that L may be considered an extension field of (a field isomorphic to) K. Then show that L is generated over this field by the coset $X + (p(X))$. Finally, show that $X + (p(X))$ is a root of p.

35.2* Generalize each definition and, with proof, each proposition and theorem in Chapter 2. (Hints: In Definition 8.2, replace "The field generated by α over K" with "A field generated by α over K" and use minimality in the definition, as in the definition of a generated group (4.1). Make a similar modification to Definition 8.4, noting that $K[\alpha]$ may be considered as a subring of the extension field L containing K and α.)

35.3* Prove Theorem 35.3. (Hint: Adapt the proof of Theorem 11.3, considering carefully the action of the derivative on terms of the form cX^{np} when the characteristic of K is p, to derive the theorem in the case where the roots are all distinct. Note that the greatest common divisor is defined only for *nonzero* polynomials. Now for the other case, observe that if $(X - \alpha)$ divides $f(X)$, then $(X - \alpha^p)$ divides $g(X)$. Show that a factorization for g yields a factorization for f with factors of the form $X^p - \beta$. Then show that if α is the root of some factor $X^p - \beta$, then $X^p - \beta = (X - \alpha)^p$.)

35.4* Prove that if either (a) K is of characteristic 0 or (b) K is of characteristic p and K^p, the set of all pth powers of elements of K, is equal to K itself, then all polynomials in $K[X]$ are separable. (Hint: For (b), consider Exercise 35.3 and ask whether $X^p - \beta$, and hence an f of the form $g(X^p)$ that factors into linear terms in an extension field L, may ever be irreducible over K.) Fields satisfying condition (a) or (b) are said to be *perfect*.

35.5* Let $\mathbb{F}_p = \mathbb{Z}/p\mathbb{Z}$ be the finite field with p elements, and let $K = \mathbb{F}_p(t)$ be the field of quotients of polynomials in one indeterminate t having coefficients in $\mathbb{F}_p$. (See Example 19.4 for a related field.) Consider $f(X) = X^p - t$. Show that $f \in K[X]$ is irreducible and inseparable. (Hint: Use Theorem 35.3 for inseparability. For irreducibility, suppose a factorization $f = gh$ into nonconstant monic polynomials g and h exists over K. Let α be a root of f in a splitting field L of f over K. Then, in L, both g and h are of the form $(X - \alpha)^m$, $1 \le m < p$. But then the constant term of g, say α^m, lies in K. Choose an appropriate power of α^m to show that $\alpha \in K$. But then α is a quotient $a(t)/b(t)$ of polynomials in t, hence $a(t)^p - tb(t)^p = 0$. Consider highest terms of $a(t)$ and $b(t)$ to derive a contradiction.)

35.6* Prove that all finite fields are perfect. (Hint: A finite field K has positive characteristic p. Consider the *Frobenius* map $x \mapsto x^p$ on K. Show that it is a ring homomorphism. Is it one-to-one? Onto?)

35.7* Adding additional hypotheses on separability where necessary, generalize each definition, and, with proof, each corollary and theorem in Chapter 3. (Hint: For Theorem 11.2, either use an algebraic closure of the field, replace linear factors with irreducibles, or proceed as in Exercise 13.10. For Theorem 12.12, alter the hypotheses to start with an algebraic extension K/F and an algebraic element α over K in some extension field L of K; show that every element of $K(\alpha)$ is algebraic over F.)

35.8* Prove that given a field K and a polynomial p, there exists a field extension L of K such that p factors into linear terms over L. (Hint: First assume that p is irreducible. Then use Kronecker's Theorem (35.1) inductively.)

35.9* Prove that any two splitting fields of a separable polynomial $p \in K[X]$ over a field K are isomorphic over K. (Hint: Adapt the theorems of section 18, finally using a generalization of the Extension of Isomorphism to Splitting Fields Theorem (18.7).)

35.10* Generalize Theorem 35.7 to an extension L/K generated by the roots of arbitrarily many separable polynomials in $K[X]$, paralleling Theorem 19.1.

35.11* Prove that if K is a finite field, then its multiplicative group K^* is cyclic. (Hint: Let e be the exponent of the abelian group K^*. Show (a) that $e \le |K^*|$ using the Fundamental Theorem of Finite Abelian Groups; (b) that every element of K^* is a root of $X^e - 1$; (c) that, by Theorem 11.1, there are at most e roots of $X^e - 1$ in K; (d) that $e = |K^*|$; and finally, (e) that K^* is cyclic, using the Fundamental Theorem again.)

35.12* Adding additional hypotheses on separability where necessary, generalize each definition and, with proof, each corollary and theorem in Chapter 4. (Hint: For Theorem 16.2, first assume that K is an infinite field and use Theorem 35.7 to show that separability is retained. In the case where K is finite, show that $L = K(\alpha, \beta)$ is finite and use Exercise 35.11 to show that $L = K(\gamma)$ for a generator γ of the cyclic group L^*.) A related result, sometimes called the Theorem of the Primitive Element, is that a field extension is simple if and only if it has finitely many intermediate fields; see [21, Thm. I.5.6] or [24, Thm. 65 and Cor. 67].

35.13* Let $\tau : K \to K'$ be an isomorphism of fields and $p \in K[X]$ a separable polynomial. Show that $\tau(p)$ is separable over K'. (Hint: Show first that it is sufficient to handle only fields L in the definition of separability that are generated by the roots of p, that is, splitting fields L of p over K. Then use the Extension of Isomorphism to Splitting Fields Theorem (18.7)).

35.14* Adding additional hypotheses on separability where necessary, generalize each definition and, with proof, each corollary, proposition, and theorem in Chapter 5. Note how Exercises 31.5 and 31.6 assist in generalizing Theorem 24.10.

36. Finite Fields

Finite fields are fields with a finite number of elements, and the theory of finite fields is one of the first applications of the Galois theory of arbitrary fields. However, it is not necessary that you have studied the details of the proofs of the results in the previous section, so long as you assume the basic contours of the existence of splitting fields over arbitrary fields.

The simplest examples of finite fields are given by the integers modulo p for different primes p; these fields are denoted $\mathbb{Z}/p\mathbb{Z}$ or $\mathbb{F}_p$. However, we have other examples of finite fields of order p^n, $n > 1$, for primes p:

Example 36.1. Start with $\mathbb{F}_2$ as a base field, let $f(X) = X^2 + X + 1 \in \mathbb{F}_2[X]$, and form the quotient ring $K = \mathbb{F}_2[X]/(f(X))$. Note that $\mathbb{F}_2$ lies inside K as $\{0 + (f(X)), 1 + (f(X))\}$.

Now let α be the coset $X + (f(X))$. Using the Division Algorithm, check that every coset may be represented by $p + (f(X))$ where $p \in \mathbb{F}_2[X]$ and either $p = 0$ or $\deg p < 2$: each such $p = c_0 + c_1 \alpha$ for some $c_0, c_1 \in \mathbb{F}_2$. Furthermore, all such choices of c_0, c_1 yield distinct cosets. Hence K has four elements.

Check that the multiplicative inverse of α is $\alpha + 1 = X + 1 + (f(X))$:

$$\alpha(\alpha + 1) = X^2 + X + (f(X)) = -1 + (f(X)) = 1 + (f(X)).$$

Similarly, the multiplicative inverse of 1 is itself. As a result, K is a field, and we see that K is generated over $\mathbb{F}_2$ by α: $K = \mathbb{F}_2(\alpha)$.

Example 36.2. Now start with $\mathbb{F}_5$, let $f(X) = X^3 + X + 1 \in \mathbb{F}_5[X]$, and form the quotient ring $K = \mathbb{F}_5[X]/(f(X))$. As in the last example, K is a field, generated over $\mathbb{F}_5$ by $\alpha = X + (f(X))$, and has $5^3 = 125$ elements. Equivalently, $K = \mathbb{F}_5(\alpha)$ and $[K : \mathbb{F}_5] = 3$.

These examples fall into a general type, following Kronecker's Theorem (35.1). Given a prime p and a polynomial $f \in \mathbb{F}_p[X]$ that is irreducible and of degree n, then $L = \mathbb{F}_p[X]/(f(X))$ is a finite field of p^n elements, as follows. First, since $f(X)$ is irreducible, the principal ideal $(f(X))$ is maximal and so L is a field. Then, $1 + (f(X))$ generates an additive subgroup of L isomorphic to $\mathbb{Z}/p\mathbb{Z}$; hence $\mathbb{F}_p$ naturally lies in L (see Exercise 5.7). Finally, since $L/\mathbb{F}_p$ is a field extension, L is a vector space over $\mathbb{F}_p$, with basis $\{1 + (f(X)), X + (f(X)), \ldots, X^{n-1} + (f(X))\}$. Hence $[L : \mathbb{F}_p] = n$ and $|L| = p^n$.

Now note that every finite field L must have a prime subfield isomorphic to $\mathbb{F}_p$ for some prime p and that such a field L is then a vector space over $\mathbb{F}_p$. We deduce that every finite field has cardinality p^n for some prime p and natural number n.

At this point, we do not yet know that for every prime p and natural number n, there exists a polynomial in $\mathbb{F}_p[X]$ of degree n and irreducible over $\mathbb{F}_p$, which we need in order to construct a finite field of order p^n as above. We will show that, in fact, not only do such polynomials exist but that any two finite fields of the same order p^n are isomorphic. We will do so, however, in a roundabout way.

Instead of viewing finite fields as quotient fields of polynomial rings, we will consider them first as splitting fields of polynomials – not splitting fields of irreducible polynomials, but of *reducible* polynomials. The reducible polynomials we choose will take a particularly simple form: $X^q - X$ for $q = p^n$. This approach allows us to work with the finite fields rather explicitly (for every element of the field will be a root of such a polynomial!), and later we will deduce that irreducible polynomials $\mathbb{F}_p[X]$ of every degree exist.

First let us examine splitting fields of these particular polynomials:

Proposition 36.3. *Let p be a prime and $q = p^n$. The splitting field L of $f(X) = X^q - X$ over $\mathbb{F}_p$ has q elements, and every one of these elements is a root of $f(X)$.*

Proof. First we show that our polynomial $f(X)$ has distinct roots in L, so that L has at least q elements. We adapt the idea of the proof of Theorem 11.3, with K replaced by $\mathbb{F}_p$, p replaced by $f(X) = X^q - X$, and the irreducibility of p replaced by the fact that $f'(X) = -1$.

Let $\{\alpha_i\}$ be the set of roots of f in L, and suppose that the multiplicity of some root α_i is at least two. Consider the derivative $f'(X) \in \mathbb{F}_p[X]$ in two ways. First, a direct computation shows that $f'(X) = qX^{q-1} - 1$. Since $q = 0 \in \mathbb{F}_p$, $f'(X) = -1$.

On the other hand, since α_i has multiplicity at least two in f, then $f = (X - \alpha_i)^2 g$ for some polynomial $g \in L[X]$, and then

$$f'(X) = 2(X - \alpha_i)g(X) + (X - \alpha_i)^2 g'(X) = (X - \alpha_i)h(X)$$

for some polynomial $h(X) = 2g(X) + (X - \alpha_i)g'(X) \in L[X]$. But if $h(X) = 0$, then $f'(X) = 0$, a contradiction, while if $h(X) \neq 0$, then $\deg f'(X) \geq 1$, also a contradiction. Hence $f(X)$ has distinct roots in L.

To show that L contains no more than q elements, we will prove the last statement of the proposition, that *every* element of L is a root of $f(X)$. Since $f(X)$ cannot have more than $q = \deg f$ elements in any field (Theorem 11.1), we will be done.

Observe first that the elements of the prime subfield, $0, 1, \ldots, p - 1$, are all roots of $f(X)$. (For the nonzero elements, use Lagrange's Theorem for the multiplicative group $\mathbb{F}_p^*$.) Now suppose that α and β are any roots of $f(X)$ in L. Then $\alpha^q = \alpha^q$ and $\beta^q = \beta$, and moreover, since the characteristic of L is p, we have $(\alpha + \beta)^q = \alpha^q + \beta^q$ (Exercise 36.1). Then

$$(\alpha + \beta)^q - (\alpha + \beta) = (\alpha^q + \beta^q) - (\alpha + \beta) = 0$$

and so $\alpha + \beta$ is a root of $f(X)$. A direct calculation with $f(X)$ shows that $\alpha\beta$ and α^{-1} are also roots of $f(X)$. Hence every arithmetic combination of $\mathbb{F}_p$ and the roots of $f(X)$ – that is, every element of the field L generated over $\mathbb{F}_p$ by the roots of $f(X)$ – is a root of $f(X)$. □

Now we see that in fact *every* finite field is one of these splitting fields:

Proposition 36.4. *Let L be a finite field with q elements, and let p be the characteristic of L. Then $q = p^n$ for some n, and L is a splitting field for $f(X) = X^q - X$ over $\mathbb{F}_p$.*

Proof. The field L is a finite abelian group under addition and is therefore isomorphic to a direct sum of cyclic groups of prime-power order, by the Fundamental Theorem of Finite Abelian Groups. If the order of any summand were not p, the characteristic of the field would not be p. Hence L is a direct sum of cyclic groups of order p – and then $q = p^n$.

The set L^* of nonzero elements of L is of order $q - 1$, and it is a group under multiplication. By Lagrange's Theorem, $l^{q-1} = 1$ and hence $l^q = l$ for all $l \in L^*$. Since $0^q = 0$ as well, $f(X) = X^q - X = 0$ has every element of L as a root. Then, since $f(X)$ cannot have more than q roots in any extension field (Theorem 11.1), $f(X)$ factors into linear factors over L. Since the characteristic of L is p, the prime subfield of L is $\mathbb{F}_p$ (Exercise 5.7). Hence L contains a splitting field L' of $f(X)$ over $\mathbb{F}_p$.

Suppose that $L' \neq L$. Then since $|L'| < |L| = q$, there must exist a root, say α, of $f(X)$ of multiplicity greater than 1. But $f(X)$ has no multiple roots in a splitting field, by the beginning of the proof of the previous result. $\qquad\square$

Now we can show that there is one and only one finite field of each order $q = p^n$, up to isomorphism.

Corollary 36.5. *Any two finite fields of the same order are isomorphic.*

Proof. Splitting fields of the same polynomial over identical fields are isomorphic (Exercise 35.9). $\qquad\square$

In light of the preceding corollary, it is traditional to denote the finite field with q elements as $GF(q)$, for *Galois field with q elements*.

A surprising but important property of finite fields is the content of the following proposition. Recall that the *exponent* of a finite group G is the minimal positive integer e such that g^e is the identity, for all $g \in G$.

Proposition 36.6. *Let L be a finite field. Then the multiplicative group L^* is cyclic.*

Proof. We know that L has $q = p^n$ elements that are all roots of $f(X) = X^q - X$ (Propositions 36.3 and 36.4). Let e be the exponent of the abelian group L^*. Now $l^q = l$ for all $l \in L^*$, hence $l^{q-1} = 1$ for all $l \in L^*$. Therefore $e \leq |L^*| = q - 1$. Now since e is the exponent, every element of L^* is a root of $X^e - 1$. By Theorem 11.1, however, there are at most e roots of

$X^e - 1$ in L. Since 0 is not a root, there are at most e roots of $X^e - 1$ in L^*, and we have $|L^*| \leq e$. Hence $e = |L^*|$. But since L^* is a finite abelian group, there exists an element l of the group L^* of order equal to the exponent e (Exercise 36.2), and consequently $L^* = \langle l \rangle$ is cyclic.   □

Now, since every finite field L is a splitting field of $X^q - X$ over its prime subfield $\mathbb{F}_p$, it is moreover a splitting field – of the same polynomial $X^q - X$ – over every intermediate subfield K. Since the polynomials $X^q - X$ have distinct roots in every splitting field (by the beginning of the proof of Proposition 36.3), we conclude that all extensions L/K of finite fields are separable. Therefore every extension L/K appears as the splitting field L over K of a separable polynomial, and so L/K is a Galois extension in the sense of section 35 and the Galois correspondence holds.

We determine the Galois groups of finite fields L over their prime subfields $\mathbb{F}_p$ in terms of a special element of the Galois group, the *Frobenius automorphism*.

Definition 36.7 (Frobenius Automorphism). Let L be a finite field with prime subfield $\mathbb{F}_p$. Then $\sigma_p : L \to L$ defined by $\sigma_p(l) = l^p$ is the *Frobenius automorphism* of L.

It is an exercise (36.4) to show that if L has $q = p^n$ elements, then, as an element of the group of automorphisms of K, σ_p has order n.

Theorem 36.8. *Let L be a finite field of order $q = p^n$. Then $\mathrm{Gal}(L/\mathbb{F}_p)$ is cyclic and is generated by σ_p.*

Proof. Consider the cyclic subgroup $S = \langle \sigma_p \rangle$ generated by σ_p. Then the subfield $\mathrm{Fix}(S)$ of elements of L fixed by every automorphism in S is identical to the subfield of elements fixed by the Frobenius automorphism, and this set is $\{l \in L \mid l^p = l\}$. But this fixed field contains every element of $\mathbb{F}_p$, and $X^p - X$ cannot contain more than p roots in L (Theorem 11.1). Hence $\mathrm{Fix}(S) = \mathbb{F}_p$ and, by the Galois correspondence, we are done.   □

It is an exercise (36.5) to prove the following corollary.

Corollary 36.9. *Let L be a finite field of order $q = p^n$ and K a subfield of order p^m. Then $\mathrm{Gal}(L/K) = \langle \sigma_p^m \rangle$.*

Now we come full circle: for every prime p and natural number n, there is an irreducible polynomial in $\mathbb{F}_p[X]$ of degree n. To see this, let $q = p^n$ and take the splitting field L of $f(X) = X^q - X$. We know that $[L : \mathbb{F}_p] = n$, and moreover $L = \mathbb{F}_p(\alpha)$ for some $\alpha \in L$ (Exercise 36.3). Consider the minimal polynomial $m_{\alpha, \mathbb{F}_p}(X)$, which is irreducible. By the generalization of Theorem 12.9 to arbitrary fields, $\deg m_{\alpha, \mathbb{F}_p} = n$. (Alternatively, the generalization to arbitrary fields of the Structure Theorem for Fields Generated by an Algebraic Number (8.12) shows that $L \cong L' = \mathbb{F}_p[X]/(m_{\alpha, \mathbb{F}_p}(X))$. But our argument after the examples at the beginning of this section showed that $[L' : \mathbb{F}_p] = \deg m_{\alpha, \mathbb{F}_p}$. Hence $\deg m_{\alpha, \mathbb{F}_p} = n$.) Therefore $m_{\alpha, \mathbb{F}_p}$ is an irreducible polynomial over $\mathbb{F}_p$ of degree n.

Exercises

36.1* Using the unique factorization of natural numbers into products of primes, show that the binomial coefficients $\binom{p}{i}$ for $i = 1, \ldots, p-1$ are all divisible by p. Deduce that $(a + b)^p = a^p + b^p$ if a and b lie in a field of characteristic p. Then, by induction, show that furthermore

$$(a + b)^{p^n} = a^{p^n} + b^{p^n}$$

for all n.

36.2* The Fundamental Theorem of Finite Abelian Groups tells us that every finite abelian group is isomorphic to a direct sum of cyclic groups of prime-power order. Use this fact to show that the exponent of a finite abelian group is equal to the maximum order of elements in the group.

36.3* Use Proposition 36.6 to show that every extension L/K of finite fields is simple.

36.4* Prove that for a finite field L of order $q = p^n$, the Frobenius automorphism $\sigma_p: L \to L$ defined by $\sigma_p(l) = l^p$ has order n in the group of automorphisms of L over K. (Hint: Show that if m is the order of σ_p, then $X^{p^m} - X$ has every element of L as a root.)

36.5* Prove Corollary 36.9.

36.6 Let L/K be an extension of finite fields. By Exercise 36.3, $L = K(\alpha)$ for some algebraic element α. Let L be of order p^n and K of order p^m. Show that the set of roots of $m_{\alpha, K}$ is given by

$$\{\alpha^{p^{mk}} \mid k = 0, \ldots, n - m - 1\}.$$

36.7 Prove that $X^{p^n} - X \in \mathbb{F}_p[X]$ is the product of all monic irreducible polynomials in $\mathbb{F}_p[X]$ of degree dividing n. (Hint: First, use the beginning of the proof of Proposition 36.3

to show that the factorization of $X^{p^n} - X$ does not have repeated factors. Then let $f(X)$ be an irreducible polynomial in $\mathbb{F}_p[X]$ of degree dividing n, and consider the finite field $\mathbb{F}_p[X]/(f(X))$. Compute its order, and show that it is isomorphic to a subfield of the splitting field L of $X^{p^n} - X$ over $\mathbb{F}_p$. Then, since every element of L is a root of $X^{p^n} - X$, show that f is a factor of $X^{p^n} - X$.)

Historical Note

In what was certainly one of the great tragedies in mathematics, Evariste Galois died in a duel on May 30, 1832, at the age of twenty. He had published his first paper, "Démonstration d'un théorème sur les fractions continues périodiques" [48], while seventeen and still a student at the Lycée Louis-le-Grand in Paris. Later, while studying at the Ecole Preparatoire, he followed this note with articles on the resolution of "numerical" (that is, algebraic) equations: "Analyse d'un mémoire sur la résolution algébrique des équations" [49], "Note sur la résolution des équations numériques" [50], and "Sur la théorie des nombres" [51]. These papers, together with several important manuscripts published after his death, laid the foundation for what today we know as Galois theory.

Liouville, who edited Galois' manuscripts for publication in 1846, wrote in his introduction,

> I experienced an intense pleasure at the moment when, having filled in some slight gaps, I saw the complete correctness of the method by which Galois proves, in particular, this beautiful theorem: *In order that an irreducible equation of prime degree be solvable by radicals it is necessary and sufficient that all its roots be rational functions of any two of them* ([56]; trans., [43, pp. 376–377]).[1]

Recognizing the quality and influence of Galois' mathematical work comes fairly naturally to us. Sorting out reasons why Galois came into continued conflict with others and, consequently, his work was not seen as especially successful before his death, poses more of a challenge.

---

[1] A proof of this result requires more familiarity with group theory than we have assumed in this text, and this incommensurability provides another measure of the depth of Galois' insight. The argument turns on determining the solvable subgroups of S_p, the symmetric group on p letters, for primes p. See [1, §14.7, Ex. 20] or [6, §4.8, Ex. 11–15] for details.

Conflict, as well as disappointment, characterized much of the last years of Galois' life. His studies in secondary school at Lycée Louis-le-Grand began well; he won a prize in Latin poetry and earned a distinction in Greek translation during his first two years [45, p. 25]. In his third year, however, one of his rhetoric teachers thought him "dissipated" [47]. Agreeing with the newly appointed headmaster that Galois was insufficiently mature, at the age of fifteen, to appreciate the subtleties of a course in rhetoric, his rhetoric teachers demoted him after a trimester to repeat the preparatory course [45, pp. 29–30].

Perhaps in reaction to this turn of events, Galois developed an intense interest in mathematics that year. During the next several years, he made "marked progress" in mathematics, although his tendency was to jump to conclusions: "not enough method," his teacher M. Vernier wrote. At the same time, his other teachers increasingly found him bizarre and distracted [47]. By his fifth year, with the encouragement of his mathematics instructor, M. Richard, he set his hopes on entering the Ecole Polytechnique. He declined, however, to follow the usual preparatory course for the entrance examination. Galois failed the entrance exam twice, the second time shortly after his father, framed by political enemies, committed suicide. Galois instead entered the Ecole Preparatoire (now the Ecole Normale Supérieure), with disappointment despite the fact that its faculty had included Cauchy, Fourier, Lagrange, and Poisson [46, pp. 87–88].

Galois' mathematical research began to earn him recognition at the age of seventeen, with the publication of his paper on continued fractions [48]. He then described his most important work in three submissions to the Académie des Sciences de Paris. These *mémoires*, however, were rejected [45, pp. 41, 90]. In the last rejection, Poisson wrote, "We have made every effort to comprehend M. Galois' proof. His arguments are neither sufficiently clear nor developed for us to judge their rigor" ([46, p. 96], from [44, pp. 340–341]).

Galois' political convictions brought him into conflict, as well. Joining a republican group, the Société des Amis du Peuple, he took a stand against royalists, including Cauchy, an elder of the Académie. He was expelled from the Ecole Preparatoire for political reasons, and he spent time in the Sainte-Pélagie prison for comments during a republican banquet [46, pp. 90–92]. On parole due to an outbreak of cholera, Galois was removed to a medical clinic and then fell in love with Stéphanie Poterin du Motel, daughter of a doctor living on the same street [45, pp. 104–105]. This relationship may or may not have precipitated the duel at which he met his end; whether the duel was the result of political intrigue, an agent provocateur, or a simple feud between two friends, we will likely never know for sure. (See the bibliography for several reconstructions of Galois' life.)

Depending on how you read the story, Galois may have been a brilliant yet high-strung student misunderstood by his teachers in the humanities as well as several professional mathematicians, or an unbalanced, withdrawn genius set adrift by significant misfortunes. To what extent Galois' early disappointments, followed by his political activism and his dubiously requited romantic interest, combined to cause his early death is difficult to deduce. Galois' life remains a story of an intense young man of stunning mathematical prowess whose society, tragically, could not accommodate him.

Subgroups of Symmetric Groups

1. The Subgroups of S_4

There are eleven conjugacy classes of the 30 subgroups of S_4, as follows:

(1) $\{\langle () \rangle\}$, the class consisting of the single subgroup of order 1.

(2) The class of all subgroups of order 2 that are generated by a single transposition, that is, subgroups of the form $\langle (ij) \rangle$ for distinct i and j. This class has six elements.

(3) The class of all subgroups of order 2 that are generated by the product of two disjoint transpositions, that is, subgroups of the form $\langle (ij)(kl) \rangle$ for distinct i, j, k, and l. This class has three elements.

(4) The class of all subgroups of order 3. These are generated by a 3-cycle and take the form $\langle (ijk) \rangle$ for distinct i, j, and k. This class has four elements.

(5) The class of all cyclic subgroups of order 4. These are generated by a 4-cycle and take the form $\langle (ijkl) \rangle$ for distinct i, j, k, and l. This class has three elements.

(6) The class of all subgroups of order 4 isomorphic to the Klein 4-group $\mathbb{Z}/2\mathbb{Z} \oplus \mathbb{Z}/2\mathbb{Z}$ that are generated by two transpositions. These subgroups have the form $\langle (ij), (kl) \rangle$ for distinct i, j, k, and l. This class has three elements.

(7) $\{\langle (12)(34), (13)(24) \rangle\}$, the class consisting of the unique subgroup of order 4 isomorphic to the Klein 4-group $\mathbb{Z}/2\mathbb{Z} \oplus \mathbb{Z}/2\mathbb{Z}$ that is generated by two products of two transpositions.

(8) The class of all subgroups of order 6. These have the form $\langle (ijk), (ij) \rangle$ for distinct i, j, and k. This class has four elements.

(9) The class of all subgroups of order 8. These are isomorphic to D_4, the dihedral group of order 8. These have the form $\langle(ijkl), (ik)\rangle$ for distinct i, j, k, and l. This class has three elements.

(10) $\{A_4\}$, the class consisting of the single subgroup of order 12, A_4.

(11) $\{S_4\}$, the class consisting of the single subgroup of order 24, S_4 itself.

The nine transitive subgroups are partitioned into five conjugacy classes, namely 5, 7, 9, 10, and 11.

2. The Subgroups of S_5

There are nineteen conjugacy classes of the 156 subgroups of S_5, as follows:

(1) $\{\langle()\rangle\}$, the class consisting of the single subgroup of order 1.

(2) The class of all subgroups of order 2 that are generated by a single transposition, that is, subgroups of the form $\langle(ij)\rangle$ for distinct i and j. This class has ten elements.

(3) The class of all subgroups of order 2 that are generated by the product of two disjoint transpositions, that is, subgroups of the form $\langle(ij)(kl)\rangle$ for distinct i, j, k, and l. This class has fifteen elements.

(4) The class of all subgroups of order 3. These are generated by a 3-cycle and take the form $\langle(ijk)\rangle$ for distinct i, j, and k. This class has ten elements.

(5) The class of all cyclic subgroups of order 4. These are generated by a 4-cycle and take the form $\langle(ijkl)\rangle$ for distinct i, j, k, and l. This class has fifteen elements.

(6) The class of all subgroups of order 4 isomorphic to the Klein 4-group $\mathbb{Z}/2\mathbb{Z} \oplus \mathbb{Z}/2\mathbb{Z}$ that are generated by two transpositions. These subgroups have the form $\langle(ij), (kl)\rangle$ for distinct i, j, k, and l. This class has fifteen elements.

(7) The class of all subgroups of order 4 isomorphic to the Klein 4-group $\mathbb{Z}/2\mathbb{Z} \oplus \mathbb{Z}/2\mathbb{Z}$ that are generated by two products of two transpositions. These subgroups have the form $\langle(ij)(kl), (ik)(jl)\rangle$ for distinct i, j, k, and l. This class has five elements.

(8) The class of all cyclic subgroups of order 5. These are generated by a 5-cycle and take the form $\langle(ijklm)\rangle$ for distinct i, j, k, l, and m. This class has six elements.

(9) The class of all cyclic subgroups of order 6. These are generated by a 3-cycle and a transposition that are disjoint, and they hence take the form $\langle (ijk), (lm) \rangle$ for distinct i, j, k, l, and m. This class has ten elements.

(10) The class of all subgroups of order 6 isomorphic to S_3 that fix two of the letters from 1 to 5. These have the form $\langle (ijk), (ij) \rangle$ for distinct i, j, and k. This class has ten elements.

(11) The class of all subgroups of order 6 isomorphic to S_3 that do not fix any letter. These have the form $\langle (ijk), (ij)(lm) \rangle$ for distinct i, j, k, l, and m. This class has ten elements.

(12) The class of all subgroups of order 8 isomorphic to D_4, the dihedral group of order 8. These have the form $\langle (ijkl), (ik) \rangle$ for distinct i, j, k, and l. This class has fifteen elements.

(13) The class of all subgroups of order 10 isomorphic to D_5, the dihedral group of order 10. These have the form $\langle (ijklm), (jm)(kl) \rangle$ for distinct i, j, k, l, and m. This class has six elements.

(14) The class of all subgroups of order 12 isomorphic to A_4. These take the form $\langle (ij)(kl), (ik)(jl), (jkl) \rangle$ for distinct i, j, k, and l. This class has five elements.

(15) The class of all subgroups of order 12 isomorphic to the direct product $S_3 \times \mathbb{Z}/2\mathbb{Z}$. These take the form $\langle (ij), (ijk), (lm) \rangle$ for distinct i, j, k, l, and m. This class has ten elements.

(16) The class of all subgroups of order 20. These take the form $\langle (ijklm), (ijlk) \rangle$ for distinct i, j, k, l, and m. This class has six elements.

(17) The class of all subgroups of order 24. These are isomorphic to S_4 and each fix a single letter from 1 to 5. Hence they take the form $\langle (ijk), (ijkl) \rangle$ for distinct i, j, k, and l. This class has five elements.

(18) $\{A_5\}$, the class consisting of the single subgroup of order 60, A_5.

(19) $\{S_5\}$, the class consisting of the single subgroup of order 120, S_5 itself.

The twenty transitive subgroups are partitioned into five conjugacy classes, namely 8, 13, 16, 18, and 19.

Bibliography

Introductions to abstract algebra

[1] Dummit, David S. and Richard M. Foote. *Abstract algebra*, 2nd ed. New York: John Wiley and Sons, 1999.

[2] Fraleigh, John B. *A first course in abstract algebra*, 7th ed. Boston: Addison-Wesley, 2003.

[3] Gallian, Joseph A. *Contemporary abstract algebra*, 5th ed. Boston: Houghton Mifflin, 2002.

[4] Herstein, I. N. *Abstract algebra*, 3rd ed. Upper Saddle River, NJ: Prentice-Hall, 1996.

[5] Hungerford, Thomas W. *Abstract algebra: an introduction*, 2nd ed. Belmont, CA: Brooks/Cole, 1997.

[6] Jacobson, Nathan. *Basic algebra I*, 2nd ed. New York: W. H. Freeman and Company, 1985.

[7] Rotman, Joseph. *A first course in abstract algebra*, 2nd ed. Upper Saddle River, NJ: Prentice-Hall, 2000.

Texts in advanced algebra and Galois theory

[8] Artin, Emil. *Galois theory*. Edited by and with a supplemental chapter by Arthur N. Milgram. Notre Dame Mathematical Lectures, no. 2. Mineola, NY: Dover Publications, 1998.

[9] Ayad, Mohamed. *Théorie de Galois. 122 exercices corrigés (niveau I Licence-Maîtrise)*. Paris: Ellipses, 1997.

[10] Ayad, Mohamed. *Théorie de Galois. 115 exercices corrigés (niveau II Maîtrise-Agrégation-DEA)*. Paris: Ellipses, 1997.

[11] Bastida, Julio R. *Field extensions and Galois theory*. Encyclopedia of mathematics and its applications, vol. 22. Menlo Park, CA: Addison-Wesley, 1984.

[12] Edwards, Harold M. *Galois theory*. Graduate Texts in Mathematics, vol. 101. New York: Springer-Verlag, 1984.

[13] Escofier, Jean-Pierre. *Théorie de Galois: Cours avec exercices corrigés*. Enseignement des Mathématiques. Paris: Masson, 1997. English translation: *Galois theory*, Leila Schneps, trans. Graduate Texts in Mathematics, vol. 204. New York: Springer-Verlag, 2001.

[14] Fenrick, Maureen H. *Introduction to the Galois correspondence*, 2nd ed. Boston: Birkhäuser, 1992.

[15] Gaal, Lisl. *Classical Galois theory, with examples*, 5th ed. Providence, RI: American Mathematical Society, 1971.

[16] Garling, J. H. *A course in Galois theory*. Cambridge: Cambridge University Press, 1986.

[17] Hadlock, Charles R. *Field theory and its classical problems*. Carus Mathematical Monographs, vol. 19. Washington, DC: Mathematical Association of America, 1978.

[18] King, R. Bruce. *Beyond the quartic equation*. Boston: Birkhäuser, 1996.

[19] Jacobson, Nathan. *Basic algebra II*, 2nd ed. New York: W. H. Freeman and Company, 1989.

[20] McCarthy, Paul J. *Algebraic extensions of fields*. Waltham, MA: Blaisdell Publishing Company, 1966.

[21] Morandi, Patrick. *Field and Galois theory*. Graduate Texts in Mathematics, vol. 167. New York: Springer-Verlag, 1996.

[22] Postnikov, M. M. *Foundations of Galois theory*. Trans. Ann Swinfen. International Series of Monographs on Pure and Applied Mathematics, vol. 29. Oxford: Pergamon Press, 1962.

[23] Roman, Steven. *Field theory*. New York: Springer-Verlag, 1995.

[24] Rotman, Joseph. *Galois theory*, 2nd ed. Universitext. New York: Springer-Verlag, 1998.

[25] Stewart, Ian. *Galois theory*, 3rd ed. London: Chapman & Hall/CRC Press, 2004.

[26] Tignol, Jean-Pierre. *Galois' theory of algebraic equations*. River Edge, NJ: World Scientific, 1988.

Texts on Euclidean constructions and geometry

[27] Greenberg, Marvin J. *Euclidean and non-Euclidean geometries: development and history*, 3rd ed. New York: W. H. Freeman and Company, 1993.

[28] Martin, George E. *Geometric constructions*. Undergraduate Texts in Mathematics. New York: Springer-Verlag, 1998.

Papers and monographs on Galois theory

[29] Berndt, Bruce C., Blair K. Spearman, and Kenneth S. Williams. Commentary on an unpublished lecture by G. N. Watson on solving the quintic. *Math. Intelligencer* **24** (2002), no. 4, 15–33.

[30] Jensen, Christian, Arne Ledet, and Noriko Yui. *Generic polynomials: constructive aspects of the inverse Galois problem*. Cambridge: Cambridge University Press, 2002.

[31] Kappe, Luise-Charlotte and Bette Warren. An elementary test for the Galois group of a quartic polynomial. *Amer. Math. Monthly* **96** (1989), no. 2, 133–137.

[32] Landau, Susan. How to tangle with a nested radical. *Math. Intelligencer* **16** (1994), no. 2, 49–55.

[33] Landau, Susan. $\sqrt{2} + \sqrt{3}$: four different views. *Math. Intelligencer* **20** (1998), no. 4, 55–60.

[34] Matzat, B. H., J. McKay, and K. Yokoyama, eds. *Algorithmic methods in Galois theory*. J. Symbolic Comput. **30** (2000), no. 6. London: Academic Press, 2000, pp. 631–872.

[35] Mináč, Ján. Newton's identities once again! *Amer. Math. Monthly* **110** (2003), no. 3, 232–234.

[36]   Roth, R. L. On extensions of $\mathbb{Q}$ by square roots. *Amer. Math. Monthly* **78** (1971), no. 4, 392–393.

[37]   Stauduhar, Richard P. The determination of Galois groups. *Math. Comp.* **27** (1973), 981–996.

On factorization of polynomials

[38]   Cohen, Henri. *A course in computational algebraic number theory*. Graduate Texts in Mathematics, vol. 138. New York: Springer-Verlag, 1993.

[39]   Landau, Susan. Factoring polynomials quickly. *Notices Amer. Math. Soc.* **34** (1987), no. 1, 3–8.

[40]   Landau, Susan. Factoring polynomials over algebraic number fields. *SIAM J. Comput.* **14** (1985), no. 1, 184–195. Erratum, *SIAM J. Comput.* **20** (1991), no. 5, 998.

[41]   Lenstra, A. K., H. W. Lenstra, Jr., and L. Lovász. Factoring polynomials with rational coefficients. *Math. Ann.* **261** (1982), 515–534.

[42]   van Hoeij, Mark. Factoring polynomials and the knapsack problem. *J. Number Theory* **95** (2002), no. 2, 167–189.

Historical reconstructions of Galois' life

[43]   Bell, E. T. *Men of mathematics*. New York: Simon and Schuster, 1937.

[44]   Bertrand, Joseph. La vie d'Evariste Galois par Paul Dupuy. *Journal des savants* (Juillet 1899), 389–400. Reprinted in *Eloges Académiques, n. s.*, Paris, 1902, 329–345.

[45]   Rigatelli, Laura Toti. *Evariste Galois: 1811–1832*. Vita mathematica, vol. 11. Trans. John Denton. Basel: Birkhäuser, 1996.

[46]   Rothman, Tony. Genius and biographers: the fictionalization of Evariste Galois. *Amer. Math. Monthly* **89** (1982), no. 2, 84–106. Updated version available http://godel.ph.utexas.edu/~tonyr/galois.html.

Related historic works

[47]   Bychan, Bernard, ed. *The Evariste Galois archive*. Available http://www.galois-group.net.

[48]   Galois, E. Démonstration d'un théorème sur les fractions continues périodiques. *Annales de mathématiques pures et appliquées*, recueil périodique rédigé et publié par J. D. Gergonne **19** (April 1829), no. 10, 294–301.

[49]   Galois, E. Analyse d'un mémoire sur la résolution algébrique des équations. *Bulletin des Sciences mathématiques physiques et chimiques*, rédigé par MM. Sturm et Gaultier de Clauvry, 1re section du Bulletin universel publié par la Société pour la propagation des connaissances scientifiques et industrielles, et sous la direction de M. le Baron de Férussac **13** (April 1830, no. 55), à Paris, chez Bachelier, quai des Grands-Augustins, §138, 271–272.

[50]   Galois, E. Note sur la résolution des équations numériques. *Bulletin de Férussac* **13** (June 1830), §216, 413–414.

[51]   Galois, E. Sur la théorie des nombres. *Bulletin de Férussac* **13** (June 1830), §218, 428–435.

[52]   Galois, E. Notes sur quelques points d'analyse. *Annales de mathématiques pures et appliquées* **21** (December 1830, no. 6), 182–184.

[53]   Kronecker, L. Ein Fundamentalsatz der allgemeinen Arithmetik. *J. Reine Angew. Math.* **100** (1887), 490–510.

[54]   Lindemann, F. Über die Zahl π. *Math. Ann.* **20** (1882), 213–225.

[55]   Liouville, J. Remarques relatives 1° à des classes très-étendues de quantités dont la valeur n'est ni rationnelle ni même réductible à des irrationnelles algébriques; 2° à un passage du livre des Principes où Newton calcule l'action exercée par une sphère sur un point extérieur *et* Nouvelle démonstration d'un théorème sur les irrationnelles algébriques. *C. R. Acad. Sci. Paris, Sér. A* **18** (1844), 883–885, 910–911.

[56]   Liouville, J., ed. Œuvres mathématiques d'Evariste Galois. Includes "Mémoire sur les conditions de résolubilité des équations par radicaux" and "Des équations primitives qui sont solubles par radicaux." *Journal de mathématiques pures et appliquées* **11** (October-November 1846), 381–444.

[57]   Steinitz, E. Algebraische Theorie der Körper. *J. Reine Angew. Math.* **137** (1910), 167–309.

Index

Printed in the United States
By Bookmasters